NECESSARY CONDITIONS FOR SIMILARITY OF CERTAIN VOLTERRA INTEGRAL OPERATORS*

Stanley J. Osher
Applied Mathematics Department
Brookhaven National Laboratory
Upton, Long Island, New York

In this paper we study the similarity properties of the Volterra operators defined on $L_p[0,1]$ for $1 < p < \infty$ by:

$$Q_{a,\nu} f = \int_0^x (x-a)^{\nu}(y-a)^{1-\nu} f(y)\,dy .$$

$0 \le a \le 1$, $0 \le \nu \le 1$.

Two linear operators, S and T, acting on $L_p[0,1]$ are said to be similar if there exists a bounded, invertible, linear transformation U such that $T = U^{-1} S U$.

A Volterra operator on $L_p[0,1]$ is defined to be a bounded, linear, completely continuous operator whose spectrum consists only of the point zero and whose form is

$$T_G : L_p[0,1] \to L_p[0,1]$$

$$T_G f = \int_0^x G(x,y) f(y)\,dy$$

for some unique kernel $G(x,y)$.

Recently, a great deal of work has been done by several Russian authors on unicellularity of operators, particularly Volterra operators. An operator T on a Banach space B is said to be one-celled if the set of its invariant subspaces is totally ordered by inclusion. An example of such a unicellular Volterra operator is the operator J^n (see below) for n a

* The research which yielded these results was supported in part by an N.S.F. Cooperative Fellowship at New York University from 10/64 to 6/66 and also in part by the U.S. Atomic Energy Commission.

Received by the editors August 15, 1966

positive integer by [6].

Clearly if S is similar to T with T unicellular, then S is unicellular.

The similarity properties of the $Q_{a,\nu}$ are important because their kernels are the canonical form of Volterra kernels having one zero on the interval $0 \le x=y \le 1$ (see the next paper). Hence, the question of the invariant subspaces of a wide class of Volterra integral operators is partially answered, merely by checking the zeros of the kernel on this interval. See Section IV for more remarks on this subject.

In the next paper, which concerns sufficient conditions, we shall show that any T_G in a wide class of Volterra operators $\{T_G\}$ whose kernels $\{G(x,y)\}$ are such that $G(x,x)$ has one zero of first order at $x = b$, $0 \le b \le 1$, is similar to a unique $K_1 Q_{a,\nu}$ and to none of the $K_2 J^\alpha$ where K_1 and K_2 are complex numbers and

$$J^\alpha f = \int_0^x \frac{(x-t)^{\alpha-1}}{\Gamma(\alpha)} f(t)\, dt , \quad \alpha > 0$$

defines J^α on $L_p[0,1]$

The similarity properties of the J^α have been exhaustively investigated by Dupras [4]. c.f. also Kalisch [6], [7], [8] and Freeman [5]. KJ^α has been shown to be the unique canonical form for a certain class of Volterra operators called operators of "order α". See [6].

We note that if $a < 0$ or if $a > 1$ and for any ν then $Q_{a,\nu}$ is similar to $(\frac{1}{2}-a)J^1$ on all $L_p[0,1]$ and to no other KJ^α. See [4] Theorem 9.7.

I. PRELIMINARIES AND CHARACTERIZATION OF $Q_{a,\nu}$

<u>I.1 Definition</u> For $0 \le a \le 1$, $0 \le \nu \le 1$, K an integer, let

$$Q_{a,\nu}^{(K+)} f = \int_0^x (x-a)^{\nu} (y-a)^{1-\nu} f(y)\, dy$$

be a mapping of $L_p[0,1]$ into itself. The branch cuts of $(x-a)^{\nu}$ and $(y-a)^{1-\nu}$ are defined by

$$(p(x))^r = |p(x)|^r e^{ir \arg p(x)}$$

and $2K\pi \le \arg p(x) < 2(K+1)\pi$.

$Q_{a,\nu}^{(K-)}$ is defined in the same way except $2(K-1)\pi < \arg p(x) \le 2K\pi$.

<u>I.2 Lemma</u> $Q_{a,\nu}^{(K+)} = Q_{a,\nu}^{(\ell+)}$ and $Q_{a,\nu}^{(K-)} = Q_{a,\nu}^{(\ell-)}$ for any pair of integers K and ℓ.

<u>Proof</u> An examination of the arguments of the various kernels yields

$$Q_{a,\nu}^{(K+)} = e^{2\pi i(K-\ell)} Q_{a,\nu}^{(\ell+)}$$

$$Q_{a,\nu}^{(K-)} = e^{2\pi i(K-\ell)} Q_{a,\nu}^{(\ell-)}$$

Henceforth we shall discuss only $Q_{a,\nu}^{(0+)} = Q_{a,\nu}^{+}$ and $Q_{a,\nu}^{(0-)} = Q_{a,\nu}^{-}$.

If there is no danger of ambiguity, we may just refer to $Q_{a,\nu}$, or perhaps Q.

<u>I.3 Lemma</u> $Q_{a,\nu}^{+}$ is isometrically equivalent to $Q_{a,\nu}^{-}$ on $L_p[0,1]$.

<u>Proof</u> Let

$$P_{\nu} f = e^{2\pi i(1-\nu)} f(x), \quad \text{if } x < a.$$

$$P_{\nu} f = f(x) \quad , \quad \text{if } x \ge a.$$

$$P_{\nu}^{-1} f = e^{2\pi i\nu} f(x) \quad , \quad \text{if } x < a.$$

$$P_{\nu}^{-1} f = f \quad , \quad \text{if } x \ge a.$$

$\|P_{\nu} f\|_p = \|f\|_p$ and $P_{\nu}^{-1} Q_{a,\nu}^{-} P_{\nu} = Q_{a,\nu}^{+}$.

I.4 Definition If T_A is a Volterra operator on $L_p[0,1]$ $T_A f = \int_0^x A(x,y)f(y)dt$, then let $T_{\tilde{A}} f = \int_0^x \tilde{A}(x,y)f(y)dt$ on $Lq[0,1]$ for $\tilde{A}(x,y) = A(1-y,1-x)$ and $q = P/P-1$.

I.5 Lemma If T_A and T_B both satisfy the hypothesis of 4, then

i) $T_{\tilde{A}}$ is a Volterra operator on L_q

ii) $\|T_A\|_P = \|T_{\tilde{A}}\|_q$

iii) T_A is similar (or isometrically equivalent) to T_B on $L_p[0,1]$ iff $T_{\tilde{A}}$ is similar (or isometrically equivalent) to $T_{\tilde{B}}$ on $Lq[0,1]$.

Proof $T_A{}^* f = \int_x^1 A(y,x)f(y)dt$ on $Lq[0,1]$

$S_{1-x}\, T_A{}^* S_{1-x} = T_{\tilde{A}}$ for $S_{1-x} f = f(1-x)$ is an isometric mapping on $Lq[0,1]$ into itself such that $S_{1-x}^2 = I$.

$$S_{1-x}(T_A{}^*)^n\, S_{1-x} = T_{\tilde{A}}{}^n \quad \text{so} \quad \|T_{\tilde{A}}{}^n\|_q = \|T_A{}^{*n}\|_q = \|T_A{}^n\|_P$$

and if $T_A = u^{-1} T_B u$ then

$$T_A{}^* = u^* \; T_B{}^* \; u^{*-1}$$

and

$$T_{\tilde{A}} = S_{1-x}\; u^*\; S_{1-x}\; T_{\tilde{B}}\; S_{1-x}\; u^{*-1}\; S_{1-x}$$

I.6 Lemma

(1) If $Q^+_{a,\nu}$ is similar (or isometrically equivalent) to T_A on $L_p[0,1]$, then $-Q^+_{1-a,1-\nu}$ is similar (or isometrically equivalent) to $T_{\tilde{A}}$ on $Lq[0,1]$.

(2) $\|Q^+_{1-a,1-\nu}\|_q = \|Q^+_{a,\nu}\|_P$. The same result holds for $Q^-_{a,\nu}$.

Proof Follows from Lemmas 3, 5 and the fact that the kernel of $Q^+_{a,\nu}$ evaluated $(1-y,1-x)$ is the same as that of $-Q^-_{1-a,1-\nu}$ at (x,y).

I.7 Lemma For any positive integer n,

$$Q^n_{a,\nu} \overset{T}{=} \frac{(x-a)^{\nu}(y-a)^{1-\nu}[(x-a)^2-(y-a)^2]^{n-1}}{2^{n-1}\,\Gamma(n)}$$

Proof is trivial.

I.8 Lemma If, for any $\beta \geq 1$, we define

$$Q^{\beta}_{a,\nu} \overset{T}{=} \frac{(x-a)^{\nu}(y-a)^{1-\nu}[(x-a)^2-(y-a)^2]^{\beta-1}}{2^{\beta-1}\,\Gamma(\beta)},$$

then $Q^{\beta_1+\beta_2} = Q^{\beta_1}_{a,\nu} \cdot Q^{\beta_2}_{a,\nu}$

Proof $Q^{\beta_1}_{a,\nu} \cdot Q^{\beta_2}_{a,\nu}$ has the kernel

$$\frac{(x-a)^{\nu}(y-a)^{1-\nu}}{\Gamma(\beta_1)\Gamma(\beta_2)} \int_y^x \frac{2(t-a)[(x-a)^2-(t-a)^2]^{\beta_1-1}[(t-a)^2-(y-a)^2]^{\beta_2-1}}{2^{\beta_1+\beta_2-1}}\,dt$$

By a simple change of variables $\eta = \frac{(x-a)^2-(t-a)^2}{(x-a)^2-(y-a)^2}$ which is valid a.e.

we obtain

$$[(x-a)^2-(y-a)^2]^{\beta_1+\beta_2-1} \cdot \frac{(x-a)^{\nu}(y-a)^{1-\nu}}{2^{\beta_1+\beta_2-1}\,\Gamma(\beta_1)\Gamma(\beta_2)} \int_0^1 \eta^{\beta_2-1}(1-n)^{\beta_1-1}\,d\eta$$

$$= \frac{(x-a)^{\nu}(y-a)^{1-\nu}}{2^{\beta_1+\beta_2-1}\,\Gamma(\beta_1+\beta_2)}[(x-a)^2-(y-a)^2]^{\beta_1+\beta_2-1}$$

I.9 Lemma

$$\|T_A\|_p \leq (A)_p = \left(\int_0^1 dx \left(\int_0^x |A(x,y)|^q\,dy\right)^{\frac{p}{q}} dx\right)^{\frac{1}{p}}$$

Proof Trivial application of Holder's inequality.

I.10 Lemma If $A = u^{-1} B u$ for A, B, u, and u^{-1} all bounded linear operators on a Banach space, then $\exists\, K > 0$ such that

$$K \leq \frac{\|A^n\|}{\|B^n\|} \leq 1/K \quad \text{for all } n.$$

Further, if A and B are both quasi-nilpotent, then $\exists\, \epsilon > 0$ such that for any $\lambda > 0$

$$\epsilon \leq \frac{\|(\lambda I - A)^{-1}\|}{\|(\lambda I - B)^{-1}\|} \leq 1/\epsilon .$$

A quasi-nilpotent if $\|A^n\|^{1/n} \to 0$ as $n \to \infty$.

Proofs are trivial.

II. RELATION OF $Q_{a,\nu}$ TO KJ^{α}

II.1 Lemma

$$(\lambda I - KJ^1)^{-1} = I/\lambda + K/\lambda^2\, T_{e^{K/\lambda(x-y)}} \qquad \text{if } \lambda \neq 0$$

Proof $(\lambda I - KJ^1)^{-1} = I/\lambda + \sum_{n=1}^{\infty} \frac{K^n J^n}{\lambda^{n+1}}$

$$\sum_{n=1}^{\infty} \frac{K^n J^n}{\lambda^{n+1}} = T_{K/\lambda^2 \sum_{n=0}^{\infty} \frac{K^n (x-y)^n}{n!\,\lambda^n}} = K/\lambda^2\, T_{e^{K/\lambda(x-y)}}$$

II.2 Lemma

$$(\lambda I - KQ)^{-1} = I/\lambda + K/\lambda^2\, T_{(x-a)^{\nu}(y-a)^{1-\nu} e^{K/2\lambda[(x-a)^2-(y-a)^2]}}$$

Proof Analogous to the previous lemma.

II.3 Estimate (Kalisch [8], Equations (6) and (7))

$\forall\, e > 0$, $\exists\, M(e)$ and $N(e)$ such that if $n \geq N(e)$, then

$$\frac{M(e)}{n^e \Gamma(n+1)} \leq \|J^n\|_P \leq \frac{1}{\Gamma(n+1) P^{1/P}}$$

II.4 Estimate $\forall\ \lambda > 0$, K complex

$$|\lambda|\ \mathrm{Re}\ \lambda/K\ [e^{\mathrm{Re}\ K/\lambda}-1] \le \|\lambda I + K\ T_{e\ K/\lambda(x-y)}\|_p$$

$$\le |\lambda| + |K| \begin{cases} e^{\mathrm{Re}\ K/\lambda} & \text{if } \mathrm{Re}\ K/\lambda \ge 0 \\ 1 & \text{if } \mathrm{Re}\ K/\lambda < 0 \end{cases}$$

Proof $\|A\|_p \ge \dfrac{\|Af\|_1}{\|f\|_\infty}$

Let $f = 1$

$$[\lambda I + K\ T_{e^{K/\lambda(x-y)}}] \cdot 1 = \lambda\ e^{K/\lambda\ x}$$

$$\|\lambda\ e^{K/\lambda\ x}\|_1 = |\lambda|\ \mathrm{Re}\ \lambda/K\ [e^{\mathrm{Re}\ K/\lambda}-1]$$

$$\|\lambda I + K\ T_{e^{K/\lambda(x-y)}}\|_p \le |\lambda| + |K|\ \|T_{e^{K/\lambda(x-y)}}\|_p$$

$$\le |\lambda| + |K| \operatorname*{ess\ sup}_{0<y<x<1} |e^{K/\lambda(x-y)}|$$

II.5 Estimate For any complex K, $R \ge 1$ and $\ell_\infty = \dfrac{1}{(\max(1-a,a))^\nu}$

it is true that for $n \ge N(R,a)$

$$\frac{|K|^n \ell_\infty \max[a^{2nR+2}, |1-2a|^{nR+1}]}{2^{nR+1}\ \Gamma(nR+2)} \le \|(K\ Q^R)^n\|_p \le \frac{|K|^n \max((1-a)^{2nR-1}, a^{2nR-1})}{2^{nR-1}\ \Gamma(nR)}$$

Proof If $f = \ell_\infty (x-a)^\nu$, then $\|f\|_\infty = 1$.

$$K^n Q^{nR} f = \frac{\ell_\infty K^n}{2^{nR}\Gamma(nR+1)} (x-a)^\nu\ [x^2-2ax]^{nR}$$

$$\|K^n Q^{nR} f\|_p \ge \frac{|K|^n \ell_\infty}{2^{nR}\Gamma(nR+1)} \int_0^1 |x-a|\,|2ax-x^2|^{nR}\ dx$$

The left inequality then follows for large n.

$$\|Q^{nR}\|_p \le \operatorname*{ess\ sup}_{0<y<x<1} \left| \frac{(x-a)^\nu (y-a)^{1-\nu} [(x-a)^2-(y-a)^2]^{nR-1}}{2^{nR-1}\ \Gamma(nR)} \right|$$

II.6 Estimate

$$\ell_\infty |\lambda| \;\text{Re}\; \lambda/K \left[1-2e^{-\text{Re}\; K/2\lambda \; a^2} + e^{\text{Re}\; K/2\lambda \; [1-2a]}\right]$$

$$\le \left\|\lambda I + T_{K(x-a)^{\nu}(y-a)^{1-\nu} \; e^{K/2\lambda[(x-a)^2-(y-a)^2]}}\right\|_P$$

$$\le |\lambda| + |K| \begin{cases} e^{\text{Re}\; K/2\lambda (1-a)^2}, & \text{if Re } K/2\lambda \ge 0 \\ e^{-\text{Re}\; K/2\lambda \; a^2}, & \text{if Re } K/2\lambda < 0 \end{cases}$$

Direct analogy of Estimate 4 using $f = \ell_\infty (x-a)^{\nu}$.

II.7 Theorem Q is not similar to KJ^{α} for any complex K if $\alpha \neq 1$ and $\alpha > 0$.

Proof

(a) Let $\alpha < 1$. By Estimates 3 and 5 for $n \ge \max[N(R,a), N(e)]$

$$\|Q^n\|_P \le \frac{\max\left((1-a)^{2n-1}, a^{2n-1}\right)}{2^{n-1}\,\Gamma(n)} \le \frac{n^{1+e} \max\left((1-a)^{2n-1}, a^{2n-1}\right)}{2^{n-1}\, M(e)} \|J^n\|_P$$

but $J^n = (KJ^{\alpha})^n \cdot (1/K\; J^{1-\alpha})^n$

so
$$\frac{\|Q^n\|_P}{\|(KJ^{\alpha})^n\|_P} \le \frac{2n^{1+e}}{M(e)\max(1-a,a)} \left\| \left(\frac{\max\left((1-a)^2, a^2\right)}{2K} J^{1-\alpha} \right)^n \right\|_P \to 0$$

since $CJ^{1-\alpha}$ is quasi-nilpotent

(b) Let $\alpha > 1$.

$$\|Q^n\|_P \ge \ell_\infty \frac{\max(a^{2(n+1)}, |1-2a|^{n+1})}{2^{n+1}\,\Gamma(n+2)}$$

$$\ge \ell_\infty \frac{\max(a^{2(n+1)}, |1-2a|^{n+1})}{2^{n+1}\,(n+1)} P^{1/P} \|J^n\|_P$$

and
$$\|(KJ^{\alpha})^n\|_P \le \left\|\left(\frac{J}{K}\right)^n\right\|_P \|(KJ^{\alpha-1})^n\|_P$$

so
$$\frac{\|Q^n\|_P}{\|(KJ^{\alpha})^n\|_P} \ge \ell_\infty \frac{\max[a^2, |1-2a|]}{2(n+1) \left\| \left(\frac{2K \quad J^{\alpha-1}}{\max(a^2, |1-2a|)} \right)^n \right\|_P} \to \infty$$

because $CJ^{\alpha-1}$ is quasi-nilpotent.

II.8 Theorem If $Q_{a,\nu}$ is similar to KJ^1, then

(1) $a = 1$ and $K = -\frac{1}{2}$ or

(2) $a = 0$ and $K = \frac{1}{2}$.

Proof By Estimates 4 and 6

$$\frac{|\lambda| \ \mathrm{Re}\ \lambda/K\ [e^{\mathrm{Re}\ K/\lambda}-1]}{|\lambda| + \begin{cases} e^{\mathrm{Re}\ 1/2\lambda\,(1-a)^2}, & \text{if } \mathrm{Re}\ 1/2\lambda \ge 0 \\ e^{-\mathrm{Re}\ 1/2\lambda\ a^2}, & \text{if } \mathrm{Re}\ 1/2\lambda < 0 \end{cases}} \le \frac{\|(\lambda I-KJ)^{-1}\|_P}{\|(\lambda I-Q)^{-1}\|_P}$$

$$\le \frac{|\lambda| + |K| \begin{cases} e^{\mathrm{Re}\ K/\lambda}, & \text{if } \mathrm{Re}\ K/\lambda \ge 0 \\ 1, & \text{if } \mathrm{Re}\ K/\lambda < 0 \end{cases}}{\ell_\infty |\lambda|\ \mathrm{Re}\ \lambda[1-2\ e^{-\mathrm{Re}\ 1/2\lambda\ a^2} + e^{\mathrm{Re}\ 1/2\lambda[1-2a]}]}$$

Call the left-hand ratio $A(\lambda)$, the right-hand one $B(\lambda)$. If $Q_{a,\nu}$ is similar to KJ^1, then by Lemma I.10 $\exists\ \varepsilon > 0$ such that for all non-zero λ

$$0 < \varepsilon \le B(\lambda), \quad \text{and } A(\lambda) \le \frac{1}{\varepsilon} < \infty$$

(a) Suppose $\mathrm{Im}\ K \ne 0$, $K = K_1 + iK_2$.

$$\mathrm{Re}\ K/\lambda = \frac{K_1\lambda_1 + K_2\lambda_2}{\lambda_1^2 + \lambda_2^2} \qquad \text{Let } \lambda_1 = 0 \text{ and } K_2\lambda_2 > 0$$

$$\mathrm{Re}\ K/\lambda = K_2/\lambda_2.$$

In $A(\lambda)$, let $|\lambda_2| \to 0$, and we get a contradiction.

(b) If $K > 0$, take $\lambda < 0$.

In $B(\lambda)$ the numerator behaves like K, the denominator behaves like $e^{a^2/2|\lambda|}$. Thus similarity is impossible if $K > 0$ and $a > 0$. If $a = 0$ and $K > 0$, then in $A(\lambda)$ take $\lambda > 0$. The numerator behaves like $e^{K/\lambda}$, the denominator like $e^{1/2\lambda}$ so for similarity we need $K \le \frac{1}{2}$. In $B(\lambda)$ take $\lambda > 0$. $B(\lambda)$ behaves like $e^{K-\frac{1}{2}/\lambda}$ so we need $K \ge \frac{1}{2}$ or $K = \frac{1}{2}$.

(c) If, for K positive, $Q_{a,\nu}$ similar to $-KJ^1$ on L_p, then by

Lemma I.6 $Q_{1-a,1-\nu}$ is similar to KJ^1 on Lq or a = 1 and K = ½ and we are finished.

<u>II.9 Lemma</u> $Q_{0,\nu}$ and $-Q_{1,\nu}$ are isometrically equivalent to $\frac{1}{2} J^1$ on $L_{1/\nu}$ for $0 \le \nu \le 1$.

Let $uf = \frac{1}{2^\nu} \frac{f(\sqrt{x})}{x^{\nu/2}}$, $u^{-1}f = x^\nu 2^\nu f(x^2)$

$$Vf = \frac{f(1-\sqrt{1-x})}{(\sqrt{1-x})^\nu 2^\nu}, \quad V^{-1}f = f(2x-x^2)(1-x)^\nu 2^\nu$$

on $L_{1/\nu}$ [0,1]. u, V, u^{-1}, and V^{-1} are isometries and

$$Q_{1,\nu} = -\tfrac{1}{2} V^{-1}J^1V; \quad Q_{0,\nu} = \tfrac{1}{2} u^{-1}J^1u.$$

<u>II.10 Estimate</u> For $0 > j \ge -\nu-1/P$, $0 \le a < \frac{1}{2}$ and $j \ne -1$ there exists $N(a,\nu,j,P)$ and $C_{P,j}(N) > 0$ such that if $n \ge N$, then

$$\frac{C_{P,j}(N)(1-a)^{2n}}{2^n n^{1/P+j/2}\Gamma(n+1)} \le \|Q^n_{a,\nu}\|_P$$

<u>Proof</u> Let $f = \ell_P(x-a)^{\nu+j}$

$$\ell_P = \left[\frac{(\nu+j)P+1}{a^{(\nu+j)P+1}+(1-a)^{(\nu+j)P+1}}\right]^{1/P}, \quad \|f\|_P = 1$$

$$\|Q^n_{a,\nu} f\|_P \le \|Q^n_{a,\nu}\|_P$$

$$Q^n_{a,\nu} f = \frac{\ell_P(x-a)^\nu}{2^n\Gamma(n+1)} \left[\frac{-1(-1)^j \sum_{K=1}^{n} a^{j+2K}[x^2-2ax]^{n-k}}{\prod_{\ell=1}^{n} (1+j/2\ell)} \binom{n+j/2}{n-K} + \frac{(x-a)^{j+2n}}{\prod_{\ell=1}^{n}(1+j/2\ell)} \right]$$

$$\|Q^n_{a,\nu}f\|_P \ge \left[\int_{2a}^{1} |Q^n_{a,\nu}f|^P\, dx\right]^{1/P} \ge \left[\int_{2a}^{1} |\operatorname{Im} e^{iz}Q^n_{a,\nu}f|^P\, dx\right]^{1/P}$$

for any real z.

For $1 \ge x \ge 2a$, the first n terms have arguments $\pm \pi(j+1)$, the $(n+1)^{st}$ term has argument zero. Let $z = \mp \pi(j+1)$. Terms one to n are now real. The $(n+1)^{st}$ term has imaginary part

$$\frac{\pm [\sin(\pi j)]\ell_P(x-a)^{j+2n+\nu}}{\Gamma(n+1) \prod_{\ell=1}^{n} (1 + j/2\ell)2^n}$$

The result now follows if we notice $(1-a)^{2n} \gg a^{2n}$ for large n and

$$\lim_{n\to\infty} \frac{\prod_{\ell=1}^{n} (1 + j/2\ell)}{n^{j/2}} = C(j) .$$

<u>II.11 Estimate</u> If $P < 1/\nu$ (or $q > \frac{1}{1-\nu}$) such that $q = \frac{2K+\delta+1}{1-\nu}$ for K a non-negative integer and $0 \le \delta < 2$ with $K + \delta/2 > 0$, $0 \le a < \frac{1}{2}$, and $R > 2/\delta$ if $\delta \ne 0$ or $R = +\infty$ if $\delta = 0$ then $\exists$ $N(a,\nu,R,P)$ and $C_{P,R}(N) > 0$ such that if $n \ge N$ then $\|Q^n_{a,\nu}\|_P \le \frac{C_{P,R}(N)(1-a)^{2n}}{2^n\Gamma(n+1)n^{K/q+1/Rq}}$

<u>Proof</u> By Lemma I.9

$$\|Q^n_{a,\nu}\|_P \le (Q^n_{a,\nu})_P$$

$$\le \frac{1}{2^{n-1}\Gamma(n)}\Bigg(\int_0^a (a-x)^{\nu P}dx\left(\int_0^x (a-y)^{(1-\nu)q}dy\cdot[(y-a)^2-(x-a)^2]^{(n-1)q}dy\right)^{P/q}$$

$$+\int_a^{2a} (x-a)^{\nu P}dx\Bigg(\int_0^{2a-x} (a-y)^{(1-\nu)q}[(y-a)^2-(x-a)^2]^{(n-1)q}dy$$

$$+\int_{2a-x}^{a} (a-y)^{(1-\nu)q}[(x-a)^2-(y-a)^2]^{(n-1)q}dy$$

$$+\int_a^x (y-a)^{(1-\nu)q}[(x-a)^2-(y-a)^2]^{(n-1)q}dy\Bigg)^{P/q}$$

$$+\int_{2a}^{1} (x-a)^{\nu P}dx\Bigg(\int_{2a-x}^{a} (a-y)^{(1-\nu)q}[(x-a)^2-(y-a)^2]^{(n-1)q}dy$$

$$+\int_a^x (y-a)^{(1-\nu)q}[(x-a)^2-(y-a)^2]^{(n-1)q}dy\Bigg)^{P/q}\Bigg)^{1/P}$$

The result follows as a simple consequence of integration by parts and Holder's inequality. The important term is the one involving $\int_{2a}^{1} (x-a)^{\nu P} dx$ ....

II.12 Theorem If $1 < P < \infty$ and either $Q_{0,\nu}$ is similar to $\frac{1}{2} J^1$ on L_P or $-Q_{1,\nu}$ is similar to $\frac{1}{2} J^1$ on L_P, then $P = 1/\nu$.

Proof (a) If $P > 1/\nu$, then by Estimate 2 and 10,

$$\lim_{n\to\infty} \frac{\|(\frac{1}{2} J^1)^n\|_P}{\|Q^n_{0,\nu}\|_P} \leq \lim_{n\to\infty} \frac{\dfrac{1}{\Gamma(n+1)P^{1/P}}}{\dfrac{C_{P,j}(N)}{n^{j/2+1/P}\Gamma(n+1)}} = 0$$

if we may choose $j \ni 1/P + j/2 < 0$ or $2/P < -j < \nu + 1/P$ which is possible because $P > 1/\nu$.

(b) Suppose $P < 1/\nu$, then by Estimates 3 and 11,

$$\lim_{n\to\infty} \frac{\|Q^n_{0,\nu}\|_P}{\|(\frac{1}{2} J^1)^n\|_P} \leq \frac{C_{P,R}(N)\ n^e\ \Gamma(n+1)}{\Gamma(n+1)n^{K/q + 1/Rq}\ M(e)} = 0$$

if we may choose e and R such that $e - K/q - 1/Rq < 0$ or $0 < e < K/q + 1/Rq < \dfrac{\delta/2 + K}{q}$ which is certainly possible.

II.13 Summary Let K be any complex number and let $\alpha > 0$, $1 < P < \infty$, $0 \leq a \leq 1$, and $0 \leq \nu \leq 1$. Then $Q_{a,\nu}$ is similar to KJ^{α} on $L_P[0,1]$ iff one of the following occurs

(1) $Q_{0,\nu} = u^{-1}\frac{1}{2}J^1 u$ on $L_{1/\nu}$

(2) $Q_{1,\nu} = -V^{-1}\frac{1}{2}J^1 V$ on $L_{1/\nu}$

If u or V is an isometry, then it must be the one mentioned in 8 as a result of [7]. Theorem 1. See Section IV: Remarks.

III. RELATIONS OF $Q_{a,\nu}$ TO $K\,Q^R_{b,\mu}$

III.1 Observations Lemma III.9 gives us the result that $Q_{a,\nu} = u^{-1}K\,Q^R_{b,\mu}\,u$ on L_P when $P = 1/\nu$, $a = 0$, $b = 1$, $R = 1$, $\mu = \nu$ and $K = -1$. We shall show that such examples are rare exceptions.

III.2 Lemma $Q_{a,\nu}$ is not similar to $K\,Q^R_{b,\mu}$ if $R > 1$.

Proof Use Estimate II.5. For n large enough

$$\|Q^n_{a,\nu}\|_P \geq \frac{\ell_\infty \max[a^{2n+2}, |1-2a|^{n+1}]}{2^{n+1}\,\Gamma(n+2)} \geq \frac{\ell_\infty \|Q^n_{b,\mu}\| \max[a^{2(n+1)}, |1-2a|^{n+1}]}{4(n+1)n\max[(1-b)^{2n-1}, b^{2n-1}]}$$

$$\frac{\|(K\,Q^R_{b,\mu})^n\|_P}{\|Q^n_{a,\nu}\|_P} \leq \frac{|K|^n \|(Q^{R-1}_{b,\mu})^n\|_P \|Q^n_{b,\mu}\|_P}{\|Q^n_{a,\nu}\|_P}$$

$$\left\|\left(\frac{K\max(b^2,(1-b)^2)}{\max(a^2,|1-2a|)}\,Q^{R-1}_{b,\mu}\right)^n\right\| \frac{4n(n+1)\max(a^2,|1-2a|)}{\max(b,1-b)\ell_\infty}$$

III.3 Lemma If $Q_{a,\nu}$ is similar to $K\,Q_{b,\mu}$ for some complex K then either (1) $K = 1$ and $a = b$ or (2) $K = -1$ and $a+b = 1$.

Proof From Estimate II.6

$$\frac{\ell_\infty|\lambda|\ \mathrm{Re}\ \lambda/K\,[1-2e^{-\mathrm{Re}\ K/2\lambda}\,b^2 + e^{\mathrm{Re}\ K/2\lambda}[1-2b]]}{|\lambda| + \begin{cases} e^{\mathrm{Re}\ \frac{1}{2}\lambda(1-a)^2}, & \text{if } \mathrm{Re}\ 1/\lambda \geq 0 \\ e^{-\mathrm{Re}\ \frac{1}{2}\lambda\ a^2}, & \text{if } \mathrm{Re}\ 1/\lambda < 0 \end{cases}} \leq \frac{\|(\lambda I - KQ_{b,\mu})^{-1}\|_P}{\|(\lambda I - Q_{a,\nu})^{-1}\|_P}$$

$$\frac{\|(\lambda I - KQ_{b,\mu})^{-1}\|_P}{\|(\lambda I - Q_{a,\nu})^{-1}\|_P} \leq \frac{|\lambda|+|K| \begin{cases} e^{\mathrm{Re}\ K/2\lambda(1-b)^2}, & \text{if } \mathrm{Re}\ K/\lambda \geq 0 \\ e^{-\mathrm{Re}\ K/2\lambda\ b^2}, & \text{if } \mathrm{Re}\ K/\lambda \leq 0 \end{cases}}{\ell_\infty|\lambda|\,\mathrm{Re}\ \lambda[1-2e^{-\mathrm{Re}\frac{1}{2}\lambda}\,a^2 + e^{\mathrm{Re}\frac{1}{2}\lambda}[1-2a]]}$$

Let $C(\lambda)$ be the left-hand ratio, $D(\lambda)$ be the right. By Lemma I.10, similarity implies $\exists\ \epsilon > 0\ \ni$

$C(\lambda) \leq 1/\epsilon < \infty$ and $D(\lambda) \geq \epsilon > 0$ for all non-zero λ.

(a) If $\text{Im } K \neq 0$, $K = K_1 + iK_2$, choose $\lambda = i\lambda_2$ such that $K_2/\lambda_2 < 0$. In $C(\lambda)$ the numerator behaves like $e^{|K_2/\lambda_2| b^2/2}$ if $b > 0$ and the denominator behaves like 1. If $b = 0$, choose $\lambda = i\lambda_2 \ni K_2/\lambda_2 > 0$. The ratio $C(\lambda)$ then behaves like $e^{K_2/2\lambda_2}$. Thus K is real.

(b) If $K > 0$. In $C(\lambda)$ take $\lambda < 0$. The numerator behaves like $e^{-K/2\lambda\, b^2}$ if $b > 0$, the denominator behaves like $e^{-1/2\lambda\, a^2}$. We thus need $Kb^2 \leq a^2$ if $b > 0$. By symmetry we need $\frac{1}{K} a^2 \leq b^2$ if $a > 0$, or $K = a^2/b^2$ if a and b are both non-zero. Suppose $b = 0$. In $D(\lambda)$ let $\lambda < 0$ and we need $a = 0$. By symmetry $a = 0 \Rightarrow b = 0$. If $a = b = 0$ and $Q_{0,\nu}$ is similar to $K\, Q_{0,\mu}$ on L_p, then $Q_{1,1-\nu}$ is similar to $K\, Q_{1,1-\mu}$ on Lq or $K = 1/1 = 1$. If $Q_{1,\nu}$ is similar to $K\, Q_{b,\mu}$ on L_p, then $Q_{0,1-\nu}$ is similar to $K\, Q_{1-b,1-\mu}$ on Lq or $b = 1$ and $K = 1$. Thus if $0 < a \leq b < 1$ then

$Q_{a,\nu}$ similar to $a^2/b^2\, Q_{b,\mu}$ on $L_p \Rightarrow Q_{1-a,1-\nu}$ is similar to a^2/b^2. $Q_{1-b,1-\mu}$ on Lq or $a/b = \frac{1-a}{1-b}$ or $a = b$ and hence $K = 1$.

(c) If $K < 0$, take $\lambda < 0$ in $D(\lambda)$. If $a > 0$, we need $K/2\lambda\,(1-b)^2 \geq -1/2\lambda\, a^2$ or $-K(1-b)^2 \geq a^2$ (even if $a = 0$). By symmetry $-1/K(1-a)^2 \geq b^2$

$$\frac{(1-a)^2}{b^2} \geq -K \geq \frac{a^2}{(1-b)^2} \quad \text{or } 1 \geq a+b$$

By Lemma I.6 we need $1 \geq 1-a+1-b$ or $a+b = 1$. Hence if $b \neq 0$ or $b \neq 1$, then $K = -1$. If $b = 1$, then $a = 0$. Take $\lambda > 0$ in $D(\lambda)$ and we need $-K/2\lambda \geq \frac{1}{2\lambda}$ and $\lambda > 0$ in $C(\lambda)$ gives $-K/2\lambda \leq \frac{1}{2\lambda}$ or $K = -1$.

If $Q_{0,\nu}$ is similar to $K\, Q_{b,\mu}$ on $L_p[0,1]$ then $Q_{1,1-\nu}$ is similar to $K\, Q_{1-b,1-\mu}$ on $Lq[0,1]$ or $b = 1$ and $K = -1$.

III.4 Estimate If $0 \le a < \frac{1}{2}$ and $P = 1/\nu$, then $\exists\ N(a,P) > 0$ and $C_P(N) > 0$ such that

$$\|Q^n_{a,\nu}\|_P \le \frac{C_P(1-a)^{2n}}{2^n\Gamma(n+1)} \qquad \text{for } n \ge N$$

Proof As in Estimate II.11, use the upper bound for $(Q^n_{0,\nu})_P$. The integration is straightforward.

III.5 Estimate If $P > 1/\nu$ ($\Leftrightarrow q < 1/1-\nu$), $0 \le a < \frac{1}{2}$.

$$R > \frac{1}{q(1-\nu)+1}, \text{ then } \exists\ N\ (a,\nu,R,P) \text{ and } C_{P,R}(N)$$

such that if $n \ge N$, then

$$\|Q^n_{a,\nu}\|_P \le \frac{C_{P,R}(1-a)^{2n}}{2^n\ n^{1/P\ +\ 1/Rq}\ \Gamma(n)}$$

Proof Again use the upper bound on the $(Q^n_{a,\nu})_P$ and also Holder's inequality.

III.6 Theorem If $0 \le a \le 1$ and $a \ne \frac{1}{2}$, then $Q_{a,\nu}$ is similar to $Q_{a,\mu}$ iff $\mu = \nu$.

Proof Sufficiency is trivial. For necessity, consider the case $0 \le a < \frac{1}{2}$.

(a) If $\nu = 1/P$, then by Estimates 4 and II.10 we have

$$\lim_{n\to\infty} \frac{\|Q^n_{a,\nu}\|_P}{\|Q^n_{a,\mu}\|_P} \le \lim_{n\to\infty} C_1 n^{j/2\ +\ 1/P} = 0 \text{ if } 2/P < -j < \mu + 1/P \text{ or if } P > 1/\mu$$

Similarity $\Rightarrow$ $P \le 1/\mu$.

If $P < 1/\mu$, then by II.10 and II.11

$$\lim_{n\to\infty} \frac{\|Q^n_{a,\mu}\|_P}{\|Q^n_{a,\nu}\|_P} \leq \lim_{n\to\infty} C_2 n^{j/2 + 1/P - K/q - 1/Rq} = 0$$

if we may choose j, K, and R such that

$$2/P < -j + 2/Rq + 2K/q < 1/P + \nu + \delta/q + 2K/q = 2/P + \delta/q + \frac{2K}{q}$$

which is possible because $2K + \delta > 0$.

(b) Suppose $P < 1/\nu$, by II.10 and II.11

$$\lim_{n\to\infty} \frac{\|Q^n_{a,\nu}\|_P}{\|Q^n_{b,\mu}\|_P} \leq \lim_{n\to\infty} C_3 n^{j/2 + 1/P - K/q - 1/Rq} = 0$$

if $2/P < -j + 2K/q + 2/Rq < 1/P + \mu + 2K+\delta/q$ or if $\nu < \mu$.

Thus for similarity it is necessary that $\nu \geq \mu$ or $\frac{1}{\mu} \geq \frac{1}{\nu} > P$. By symmetry, it follows that $\mu \geq \nu$ or $\mu = \nu$.

(c) Suppose $P > 1/\nu$. We need only examine the case $P > 1/\mu$.

By Estimates II.10 and 5

$$\lim_{n\to\infty} \frac{\|Q^n_{a,\nu}\|_P}{\|Q^n_{a,\mu}\|_P} \leq \lim_{n\to\infty} C_4 n^{1 + j/2 + 1/P - 1/P - 1/Rq} \to 0$$

if $2 < -j + 2/Rq < \mu + 1/P + 1-\nu + 1/q$ or if $\nu < \mu$. Thus similarity $\Rightarrow \nu \geq \mu$ or $\frac{1}{\nu} \leq \frac{1}{\mu} < P$ and hence by symmetry, $\mu \geq \nu$ or $\mu = \nu$.

The result follows for $\frac{1}{2} < a \leq 1$ by Lemma I.6.

<u>III.7 Estimate</u> If $\frac{1}{2} < a \leq 1$ and $0 > \ell > -(1-\nu) - 1/q$ with $\ell \neq -1$, then $\exists$ $N(1-a, 1-\nu, \ell, q)$ and $C_{q,\ell}(N) > 0$ such that if $n \geq N$, then

$$\frac{C_{q,\ell}(N)\, a^{2n}}{2^n n^{1/q + \ell/2}\, \Gamma(n+1)} \leq \|Q^n_{a,\nu}\|_P$$

<u>Proof</u> Direct consequence of Lemma I.6 and Estimate II.10.

By similar reasoning, we obtain the next three estimates.

III.8 Estimates If $q < 1/1-\nu$ ($\Leftrightarrow P > 1/\nu$), $\frac{1}{2} < a \le 1$, $P = \frac{2K+1+\delta}{\nu}$ for K a non-negative integer, $0 \le \delta < 2$ with $2K + \delta > 0$, and $R > 2/\delta$ if $\delta > 0$ and $R = +\infty$ if $\delta = 0$, then

$$\exists\ N(1-a,1-\nu,R,q) \text{ and } C_{q,R}(N) > 0$$

such that if $n \ge N$, then

$$\|Q^n_{a,\nu}\|_P \le \frac{C_{q,R}(N)\ a^{2n}}{2^n\ \Gamma(n+1)\ n^{K/P + 1/RP}}$$

III.9 Estimate If $\frac{1}{2} < a \le 1$, $P = 1/\nu$ ($\Leftrightarrow q = \frac{1}{1-\nu}$), then $\exists\ N(1-a,q) > 0$ and $C_q(N) > 0$ such that if $n \ge N$, then

$$\|Q^n_{a,\nu}\|_P \le \frac{C_q(N)\ a^{2n}}{2^n \Gamma(n+1)}$$

III.10 Estimate If $\frac{1}{2} < a \le 1$, $P < 1/\nu$ ($\Leftrightarrow q > \frac{1}{1-\nu}$), $R > \frac{2}{\nu P+1}$, then $\exists\ N(1-a,1-\nu,R,q)$ and $C_{q,R}(N)$ such that if $n \ge N$, then

$$\|Q^n_{a,\nu}\|_P \le \frac{C_{q,R}(N)\ a^{2n}}{2^n\ n^{1/q + 1/RP}\ \Gamma(n)}$$

III.11 Theorem On $L_P[0,1]$, if $0 \le a < \frac{1}{2}$, and $1 < P < \infty$, then $Q_{a,\nu}$ similar to $-Q_{1-a,\mu}$ implies $\mu + \nu = 2/P$.

Proof Use Lemmas I.10, I.6, and the last eight Estimates. The result follows in much the same manner as that of Theorem III.6.

We now consider the somewhat exceptional case, $a = \frac{1}{2}$.

III.12 Estimate (a) If $P \ge 1/\nu$ and $0 > j > -1/P - \nu$ with $j \ne -1$, then $\exists\ N(\nu,j,P)$ and $C_{P,j}(N)$ such that if $n \ge N$, then

$$\|Q^n_{\frac{1}{2},\nu}\|_P \ge \frac{C_{P,j}(N)\ (\frac{1}{2})^{2n}}{2^n \Gamma(n+1) n^{j/2 + 1/P}}$$

(b) If $P < 1/\nu$ and $0 > \ell > -1/q - (1-\nu)$ with $\ell \neq -1$, then if $n \geq N(1-\nu,\ell,q)$, it is true that

$$\|Q^n_{\frac{1}{2},\nu}\|_P \geq \frac{C_{q,\ell}(N)\,(\frac{1}{2})^{2n}}{2^n\Gamma(n+1)n^{\ell/2 + 1/q}}$$

Proof Let

$$f = \ell_P(x-\tfrac{1}{2})^{j+\nu}, \quad \ell_P = \frac{[(j+\nu)P+1]^{1/P}}{(\frac{1}{2})^{j+\nu}}$$

where $0 > j > -\nu - 1/P$. Then $\|f\|_P = 1$

$$Q^n_{\frac{1}{2},\nu}f = \frac{\ell_P}{2^n\Gamma(n+1)}\ \frac{(x-\frac{1}{2})^\nu}{\prod_{\ell=1}^{n}(1+j/2\ell)}\left[-(-1)^j\sum_{K=1}^{n}(\tfrac{1}{2})^{j+2K}(x^2-x)^{n-K}\binom{n+j/2}{n-K}+(x-\tfrac{1}{2})^{j+2n}\right]$$

Again consider $|\operatorname{Im} e^{\mp i(j+1)\pi}Q^n_{a,\nu}f|$, this time on the interval $\frac{1}{2} \leq x \leq 1$. The result follows for part (a). Part (b) is a consequence of Lemma I.6.

We again used

$$\lim_{n\to\infty}\frac{n^{j/2}}{\prod_{\ell=1}^{n}(1+j/2\ell)} = C(j), \quad 0 < C(j) < \infty$$

III.13 Estimate If $P = 1/\nu$, $\exists\ N(P) > 0$ and $C_P(N) > 0$ such that if $n \geq N$, then

$$\|Q^n_{\frac{1}{2},\nu}\|_P \leq \frac{(\frac{1}{2})^{2n}\ C_P(N)}{2^n\ \Gamma(n+1)}$$

Proof Follows immediately from estimating $(Q^n_{\frac{1}{2},\nu})_P$

III.14 Estimate (a) If $P > 1/\nu$ ($\Leftrightarrow q < \frac{1}{1-\nu}$), then $\exists\ N(\nu,R,P) > 0$ and $C_{P,R}(N) > 0$ for $R > \frac{2}{q(1-\nu)+1}$ such that if $n \geq N$ then

$$\|Q^n_{\frac{1}{2},\nu}\|_P \leq (\tfrac{1}{2})^{2n} \frac{C_{P,R}(N)}{2^n\Gamma(n)n^{1/P + 1/Rq}}$$

(b) If $P < \frac{1}{\nu}$ ($\Leftrightarrow q > \frac{1}{1-\nu}$), $R > \frac{2}{\nu P+1}$, then if $n \geq N(1-\nu,R,q) > 0$ we have

$$\|Q^n_{\frac{1}{2},\nu}\|_P \leq (\tfrac{1}{2})^{2n} \frac{C_{q,R}(N)}{2^n\Gamma(n)n^{1/q + 1/RP}}$$

Proof In part (a) we estimate $(Q^n_{\frac{1}{2},\nu})_P$ using Holder's inequality. Part (b) follows immediately from Lemma I.6.

III.15 Theorem If $Q_{\frac{1}{2},\nu}$ is similar to $Q_{\frac{1}{2},\mu}$ on $L_P[0,1]$ or if $Q_{\frac{1}{2},\nu}$ is similar to $-Q_{\frac{1}{2},\nu}$ on $L_P[0,1]$, then either $\mu = \nu$ or $\mu + \nu = 2/P$ (i.e. $\exists$ 4 possibilities).

Proof Follows from Estimates 12, 13 and 14.

III.16 Summary Let K be any complex number and let (1) $0 \leq a,b,\nu,\mu \leq 1$, (2) $R \geq 1$, (3) $1 < P < \infty$. Then if we have $Q_{a,\nu}$ similar to $K\,Q^R_{b,\mu}$ on $L_P[0,1]$, we must have

(a) $R = 1$ and either $K = 1$ or $K = -1$

(b) If $K = 1$, then $a = b$, and if $a \neq \frac{1}{2}$, then $\mu = \nu$

(c) If $K = -1$, then $a + b = 1$, and if $a \neq \frac{1}{2}$, then $\mu + \nu = 2/P$

(d) If $a = b = \frac{1}{2}$, then we may have any of the four cases $K = \pm 1$ and either $\mu = \nu$ or $\mu + \nu = 2/P$.

IV. REMARKS

The question as to whether any of the following relations may hold on $L_P[0,1]$

1) $Q_{a,\nu}$ similar to $-Q_{1-a,2/P-\nu}$

2) $Q_{\frac{1}{2},\nu}$ similar to $-Q_{\frac{1}{2},\nu}$ or

3) $Q_{\frac{1}{2},\nu}$ similar to $Q_{\frac{1}{2},2/P-\nu}$ is in general left unanswered.

If $\nu = \frac{1}{2}$, then $Q_{0,\nu}$ and $-Q_{1,\nu}$ are one-celled operators and hence, for either, given any two of its invariant subspaces, one is a subset of the other. See Livsic [10]. Thus their only invariant subspaces are, (for $P = 2$),

$$L_P[b,1] \;=\; \{f \in L_P[0,1] \,|\, f(x) \equiv 0 \text{ if } x \le b\}.$$

This also follows from Lemma II.9 and Kalisch [6] and [7].

The operators $Q_{a,\nu}$ have in addition to $L_P[b,1]$ as invariant subspaces the spaces

$$\tilde{L}_P[a,\nu,\varepsilon] \;=\; \{f \in L_P[0,1]/f = (x-a)^{\nu} g(|x-a|)$$

with $g(t) \equiv 0$ if $t \ge \varepsilon$ and $\varepsilon < \min[a,1-a]$.

If the J^{α} correspond to $n \times n$ matrices of maximal order of nilpotence, the $Q_{a,\nu}$ correspond to matrices whose Jordan canonical form is

$$m_{ij} = 0 \quad \text{if } i \ne j+1, \quad i = 2,3,\ldots,n.$$

$$m_{j_0+1,j_0} = 0 \quad \text{for some } j_0$$

$$m_{i+1,i} \;=\; 1 \quad \text{elsewhere}$$

Matrices whose zeros are at m_{j_0+1,j_0} and at $m_{n-j_0+1,n-j_0}$ are isometrically equivalent. This perhaps corresponds to $Q_{a,\nu}$ and $-Q_{1-a,2/P-\nu}$.

The only invariant subspaces of these matrices are if $x = \{x_1,x_2,\ldots,x_n\}$, then

$$L_{K,\ell,j_0} = \{x \in R^n | x_1 = x_2 \ldots x_{K-1} = x_{j_0+1} = x_{j_0+2} = x_{\ell-1} = 0\}$$

We may recall that two spectral operators are similar only if their quasi-nilpotent parts are similar. Thus for K_1, K_2, and C arbitrary complex numbers, $\alpha > 0$, $0 \leq a,\nu \leq 1$ and $1 < p < \infty$.

$K_1 I + K_2 J^{\alpha}$ is similar to $I + C\, Q_{\alpha,\nu}$ on $L_p[0,1]$ iff

(1) $P = 1/\nu$; (2) $\alpha = 1$; (3) $K_1 = 1$; (4) $K_2 = \frac{1}{2}$ and $a = 0$ or $K_2 = -\frac{1}{2}$ and $a = 1$.

ACKNOWLEDGMENT

The author wishes to thank Professor J. T. Schwartz for his guidance in this work, which is based on a doctoral dissertation at New York University.

BIBLIOGRAPHY

(1) Dunford, N. A Survey of the Theory of Spectral Operators. A. M. S. Bull. Vol. 64, (1958) pp.217-274.

(2) Dunford, N. and J. T. Schwartz. Linear Operators, Part I: General Theory. Interscience, New York (1958).

(3) Dunford, N. and J. T. Schwartz. Linear Operators, Part II: Spectral Theory. Interscience, New York (1963).

(4) Dupras, A. Doctoral Thesis, New York University (1965).

(5) Freeman, J. M. Volterra Operators Similar to J: $J : f \to \int_0^x f(t)\, dt$. Trans. Amer. Math. Society (4) Vol. 16, (1965) pp. 181-192.

(6) Kalisch, G. K. On Similarity, Reducing Manifolds, and Unitary Equivalence of Certain Volterra Operators. Ann. of Math. (2), Vol. 66 (1957), pp. 481-494.

(7) Kalisch, G. K. On Isometric Equivalence of Certain Volterra Operators. Proc. A. M. S., Vol. 12 (1961), pp. 93-98.

(8) Kalisch, G. K. On Similarity Invariants of Certain Volterra Operators in L_p. Pacific J. Math. Vol. 11, (1961), pp. 247-252.

(9) Livsic, M. S. On Spectral Decomposition of Linear Non-Self Adjoint Operators. A. M. S. Trans. (2), Vol. 5, pp. 67-114 (1957).

(10) Brodski, M. S. and Livsic, M. S. Spectral Analysis of Non-Self Adjoint Operators and Intermediate Systems. A. M. S. Trans. (2) Vol. 13, pp. 265-346 (1960).

SUFFICIENT CONDITIONS FOR SIMILARITY OF CERTAIN VOLTERRA INTEGRAL OPERATORS*

by

Stanley J. Osher
Applied Mathematics Department
Brookhaven National Laboratory
Upton, Long Island, New York

It is well known to anyone familiar with the solution of Volterra integral equations of the first kind, that a Volterra operator with kernel $G(x,y)$ is characterized by the zeros and singularities of the kernel at the diagonal $x = y$. This paper will be an investigation of sufficient conditions for Volterra operators, each of whose kernel has one zero on the diagonal, to be similar to each other. The results of the preceding paper by this author have indicated that these conditions are, in a sense, necessary.

I. PRELIMINARY DISCUSSION OF VOLTERRA OPERATORS

Let $G(x,y)$ be the unique kernel defining a Volterra operator

$$T_G : L_p[0,1] \to L_p[0,1]$$

$$T_G f = \int_0^x G(x,y)f(y)\,dy$$

The case where the operator has "order" α, i.e. $G(x,y) = (x-y)^{\alpha-1}H(x,y)$, $\alpha > 0$ and $H(x,x) > 0$ has been investigated by Dupras in his doctoral thesis [3], c.f. also Kalisch [6,7,8] and Freeman [4].

<u>I.1 Definition</u> If T_G is a Volterra operator on $L_p[0,1]$ whose complex valued kernel has the following properties

* The research which yielded these results was supported in part by an N.S.F. Cooperative Fellowship at N.Y.U. from 10/64 to 6/66 and in part by the U. S. Atomic Energy Commission.

Received by the editors August 15, 1966.

i) $G(x_0,x_0) = 0$ for $0 \le x_0 \le 1$

ii) $G(x,x) < 0$ if $0 \le x < x_0$

$G(x,x) > 0$ if $x_0 < x \le 1$

iii) $[d/dx\, G(x,x)]_{x=x_0} \neq 0$

then T_G is a Volterra operator of rank 1.

Our results show, assuming global smoothness of the kernel, that the similarity properties seem to depend merely upon local behavior on the diagonal, or in some neighborhood thereof.

We shall use Friedrichs' method. This general method shows, in suitable cases, that the perturbations of a linear operator are similar to it. The delicate part of this method involves the solution of a commutator equation. The method works particularly well with Volterra integral operators.

We choose a particular set of canonical operators of rank 1. These are analogous to the operators J^α used by Dupras [3], where

$$J^\alpha f = \int_0^x \frac{(x-y)^{\alpha-1}}{\Gamma(\alpha)} f(y)dy \qquad \forall\, \alpha > 0$$

The canonical operators in question are

$$Q_{a,\nu} = T_{(x-a)^\nu (y-a)^{1-\nu}}, \qquad 0 \le a, \nu \le 1 .$$

We shall sometimes call a particular $Q_{a,\nu} = Q$ when there is no danger of ambiguity. A detailed description of the $Q_{a,\nu}$ is found in the author's preceding paper.

Let K be a complex non-zero constant, and B a smooth kernel, in a sense to be made precise later. To show $KQ + T_B$ is similar to KQ we shall use Friedrich's method, which in this case reduces to the following procedure:

We must show $\exists\, u \ni$

(1) $$(KQ + T_B)\,u = u\,K\,Q$$

We assume $u = I - T_X/K$, where T_X is a Volterra integral operator. Then to satisfy (1) we must solve the equation

(2) $$Q\,T_X - T_X Q = T_B - \frac{T_B T_X}{K}$$

This last equation is equivalent to an integral equation

(2') $$\int_y^x [(x-a)^{\nu}(t-a)^{1-\nu}X(t,y) - X(x,t)(t-a)^{\nu}(y-a)^{1-\nu}]\,dt = B(x,y) - \frac{1}{K}\int_y^x B(x,t)X(t,y)\,dt$$

Now we use either procedure (A) or (B)

(A) (1) Let $\mu_{\alpha,a}$ be a certain linear space of kernels $A(x,y)$.

(2) For each $A \in \mu_{\alpha,a}$, let $X = \Gamma(A)$ be such that $[Q, T_{\Gamma(A)}] = T_A$ and be such that $T_X = T_{\Gamma(A)}$ is a Volterra integral operator defined on each Lebesgue space $L_p[0,1]$ for p $1 \le p \le \infty$.

(3) Solve $A + \frac{B*\Gamma(A)}{K} = B = A + 1/K\,\tilde{T}_B(A)$ for A in $\mu_{\alpha,a}$ if B belongs to a certain linear subspace $S \subset \mu_{\alpha,a}$ by demonstrating that $\exists\, n_0 \ni \tilde{T}_B^{\,n_0}(B) = 0$ and $\tilde{T}_B^{\,j}(B) \in \mu_{\alpha,a}$ for any j. Hence we may let

$$A_B = B - \frac{\tilde{T}_B(B)}{K} + \dots (-1)^{n_0-1}\frac{\tilde{T}_B^{\,n_0-1}(B)}{K^{n_0-1}}$$

and $KQ + T_B$ is similar to $KQ\ \forall\, B \in S$.

(B) (1) Define a Banach subspace $L_{\alpha,a}$ of kernels $A(x,y)$

(2) $\forall\, A \in L_{\alpha,a}$, let $X = \Gamma(A)$ which satisfies $[Q, T_{\Gamma(A)}] = T_A$ and T_X is a Volterra integral operator on all $L_p[0,1]$, $1 \le p \le \infty$.

(3) $\forall\, B \in L_{\alpha,a}$, show $\tilde{T}_B(A) = B * \Gamma(A)$ is a bounded linear mapping of $L_{\alpha,a} \to L_{\alpha,a}$ with

$$\|\tilde{T}_B(A)\| \le C\|B\|\,\|A\|$$

(4) Thus $A + 1/K\, B * \Gamma(A) = B$ has the unique solution $A_B \in L_{\alpha,a}$ if $\|B\| < |K|/C$, $A_B = [I + 1/K\, \tilde{T}_B]^{-1} B$.

Hence $T_{\Gamma(A_B)}$ is a Volterra integral operator and $KQ + T_B$ is similar to KQ for any B in $L_{\alpha,a}$ with $\|B\| < |K|/C$.

The choice of $Q_{a,\nu}$, the canonical form of rank one, has been made to facilitate the solution of the integro-commutator equation.

We then enlarge the class of operators similar to Q by means of elementary similarity transformations. These transformations reduce certain Volterra integral operators of rank one to perturbations of $KQ_{a,\nu}$. i.e. Let T_A be of rank 1, and W be the elementary transformation.

Then $W^{-1} T_A W = KQ + T_B$ will be similar to KQ if T_A satisfies certain smoothness conditions.

<u>I.2 Definition</u> Let $D = \{(x,y) \mid 1 > x > y > 0\}$. Then if T_G is a Volterra integral operator, its kernel G is defined on D.

<u>I.3 Definition</u> $T_G T_H = T_{G*H}$ exists under wide conditions on G and H and

$$G*H = \int_y^x G(x,t)H(t,y)\,dt$$

<u>I.4 Definition</u> A bounded linear operator T on a Banach space is quasi-nilpotent when

$$\lim_{n\to\infty} \|T^n\|^{1/n} = 0$$

The spectral radius of a Volterra operator is zero and hence any T_G is quasi-nilpotent iff it is a Volterra integral operator.

<u>I.5 Definition</u> Given a real number $\alpha > 0$, and a complex-valued kernel G(x,y), measurable on D, we define

$$\|G\|_{\alpha,\infty} = \operatorname*{ess\,sup}_{(x,y)\in D} |(x-y)^{1-\alpha} G(x,y)|$$

$\|G\|_{\alpha,\infty}$ is clearly a norm if we identify kernels which differ on a set of measure zero. Furthermore, if $\beta \le \alpha$, then $\|G\|_{\beta,\infty} \le \|G\|_{\alpha,\infty}$.

I.6 Lemma (Dupras [3], Lemma 2.7). If $\alpha > 0$, $\beta > 0$, $\|G\|_{\alpha,\infty} < \infty$, $\|H\|_{\beta,\infty} < \infty$, then $\|G*H\|_{\alpha+\beta,\infty} \le \beta(\alpha,\beta)\, \|G\|_{\alpha,\infty}\, \|H\|_{\beta,\infty}$ where $\beta(\alpha,\beta)$ is the Beta function.

I.7 Lemma (Dupras [3], Theorems 2.5 and 2.6). If $\|G\|_{\alpha,\infty} < \infty$ for some $\alpha > 0$, then the integral operator T_G is a bounded, quasi-nilpotent linear map of $L_p[0,1] \to L_p[0,1]$ for all p, $1 \le p \le \infty$ and $\|T_G\|_p \le 1/\alpha \|G\|_{\alpha,\infty}$.

The reader is urged to review I.1 to I.8 of the preceding paper for a discussion of the important properties of the $Q_{a,\nu}$ and of $T_{\tilde{A}}$. In particular, we shall use I.5 and I.6 for $p = 1$ and $p = \infty$ also, as the proof is simple.

II. SOLUTION OF THE COMMUTATOR EQUATION

II.1 Definition Let

$$D_{22}^{(a)} = \{(x,y) \in D \mid a \le y < x < 1\}$$
$$D_{21}^{(a)} = \{(x,y) \in D \mid \max(2a-x,0) \le y < a\}$$
$$D_{12}^{(a)} = \{(x,y) \in D \mid a \le x < \min(2a-y,1)\}$$
$$D_{11}^{(a)} = \{(x,y) \in D \mid 0 < y < x < a\}$$

This division of our domain D is important in what follows.

If there is no ambiguity involved in doing so, we shall often call $D_{ij}^{(a)} = D_{ij}$.

II.2 Proposition

$$(1)\quad \Gamma(A) = \frac{1}{(x-a)^{1-\nu}(y-a)^{\nu}} \frac{\partial^2}{\partial x \partial y} \int_a^x (\eta-a)^{1-\nu} d\eta \, \frac{A\left(\eta, a-\sqrt{(y-a)^2+(\eta-a)^2-(x-a)^2}\right)d\eta}{\left(-\sqrt{(y-a)^2+(\eta-a)^2-(x-a)^2}\right)^{1-\nu}}$$

on $D_{11} \cup D_{12}$.

$$(2)\quad \Gamma(A) = \frac{1}{(x-a)^{1-\nu}(y-a)^{\nu}} \frac{\partial^2}{\partial x \partial y} \int_a^y (\xi-a)^{\nu} d\xi \, \frac{A\left(a+\sqrt{(\xi-a)^2+(x-a)^2-(y-a)^2}, \xi\right)d\xi}{\left(\sqrt{(\xi-a)^2+(x-a)^2-(y-a)^2}\right)^{\nu}}$$

in $D_{22} \cup D_{21}$.

The branch cuts of $(x-a)^{1-\nu}$, etc., have the same definitions as those of either $Q^+_{a,\nu}$ or $Q^-_{a,\nu}$ (see preceding paper, Definition I.1). That is $\Gamma^+(A)$ corresponds to $Q^+_{a,\nu}$, $\Gamma^-(A)$ corresponds to $Q^-_{a,\nu}$. The radicands will be shown to be always positive, so $\sqrt{f} = |f|^{\frac{1}{2}}$.

<u>II.3 Lemma</u> If we call $\Gamma_{(ij)}(A)$ the restriction of $\Gamma(A)$ to D_{ij}, then $\Gamma_{(ij)}(A) = \Gamma_{(ij)}(A_{(ij)})$, where $A_{(ij)}$ is the restriction of A to D_{ij}. That is, in $D_{ij}^{(a)}$, $\Gamma(A)$ depends upon the values of A only in $D_{ij}^{(a)}$.

<u>Proof</u> (a) In D_{22}, we need

$$1 \geq a + \sqrt{(\xi-a)^2+(x-a)^2-(y-a)^2} \geq \xi \geq a \ .$$

for $a \leq \xi \leq y \leq x \leq 1$. Now $(x-a)^2 \geq (y-a)^2$, so the radicand is non-negative; also

$$1 \geq x \geq a + \sqrt{(\xi-a)^2+(x-a)^2-(y-a)^2} \geq \xi$$

since

$$(y-a)^2 \geq (\xi-a)^2$$

and

$$(x-a)^2 \geq (y-a)^2$$

(b) In D_{21}, we need

$$1 \geq a+\sqrt{(\xi-a)^2+(x-a)^2-(y-a)^2} \geq a \geq \xi \geq \max\left(0,\ a-\sqrt{(\xi-a)^2+(x-a)^2-(y-a)^2}\right)$$

for $x \geq a \geq \xi \geq y \geq \max(2a-x,0)$.

Now $(x-a) \geq (a-y) \geq 0$ so $(x-a)^2 \geq (a-y)^2$ and the radicand is non-negative. This also takes care of the remaining inequalities if we note $(y-a)^2 \geq (\xi-a)^2$.

(c) In D_{12}, we need

$$\min\left(1,\ a+\sqrt{(y-a)^2+(\eta-a)^2-(x-a)^2}\right) \geq \eta \geq a-\sqrt{(y-a)^2+(\eta-a)^2-(x-a)^2} \geq 0$$

whenever $0 \leq y \leq a \leq \eta \leq x \leq \min(2a-y,1)$ we have $a-y \geq x-a \geq 0$ or $(a-y)^2 \geq (x-a)^2$ and the radicand is thus non-negative. Also $(x-a)^2 \geq (\eta-a)^2$ and thus the inequalities are obeyed.

(d) In D_{11}, we need

$$a \ge \eta \ge a - \sqrt{(y-a)^2+(\eta-a)^2-(x-a)^2} \ge 0$$

when $0 \le y \le x \le \eta \le a$. Again $(y-a)^2 \ge (x-a)^2$ so the radicand is non-negative and $(\eta-a)^2 \le (x-a)^2$ so the remaining inequalities are valid.

<u>II.4 Definition</u> If A has the following properties, we say $A(x,y)$ is of class $\mu_{\alpha,a}$. A is defined on D such that

1) $A(x,y)$ and $A_x(x,y)$ are absolutely continuous with respect to x a.e. in $D_{22} \cup D_{21}$.

2) $A(x,y)$ and $A_y(x,y)$ are absolutely continuous w.r. to y a.e. in $D_{11} \cup D_{12}$.

3) $A(x,y)$, $A_y(x,y)$, and $A_x(x,y)$ are continuous functions of x and y simultaneously in neighborhoods of the line $y = 2a-x$ contained in D if $a \ne 0$ and $a \ne 1$. (In this and subsequent statements, the neighborhoods in question may be different for the different functions).

4) (a) If $x < a$, then $\lim_{y\to x} A(x,y) = \lim_{y\to x} A_2(x,y) = 0$ a.e.

(b) If $y > a$, then $\lim_{x\to y} A(x,y) = \lim_{x\to y} A_1(x,y) = 0$ a.e.

5) $A_{11}(x,y)$ is measurable on $D_{22} \cup D_{21}$

$A_{22}(x,y)$ is measurable on $D_{11} \cup D_{12}$

and $\exists\ \alpha > 0$ such that $|A|^{\alpha,a} < \infty$, where $|A|^{\alpha,a}$ is defined as follows

a) $\|A_{22}\|_{\alpha,a,D_{11}} = \operatorname{ess\,sup}_{D_{11}} |(x-y)^{1-\alpha}A_{22}(x,y)| < \infty$

b) $\|A_{22}\|_{\alpha,a,D_{12}} = \operatorname{ess\,sup}_{D_{12}} |(x-y)^{1-\alpha}A_{22}(x,y)| < \infty$

c) $\|A_{11}\|_{\alpha,a,D_{21}} = \operatorname{ess\,sup}_{D_{21}} |(x-y)^{1-\alpha}A_{11}(x,y)| < \infty$

d) $\|A_{11}\|_{\alpha,a,D_{22}} = \operatorname{ess\,sup}_{D_{22}} |(x-y)^{1-\alpha}A_{11}(x,y)| < \infty$

e) $\|A_1\|^{\alpha,a} = \operatorname{ess\,sup}_{\max(0,2a-1)\le y\le a} |(a-y)^{-\alpha}A_1(2a-y,y)| < \infty$

f) $\|A_2\|^{\alpha,a} = \operatorname*{ess\,sup}_{\max(0,2a-1)\le y\le a} |(a-y)^{-\alpha} A_2(2a-y,y)| < \infty$

g) $\|A\|^{1+\alpha,a} = \operatorname*{ess\,sup}_{\max(0,2a-1)\le y\le a} |(a-y)^{1-\alpha} A(2a-y,y)| < \infty$

then

$$|A|^{\alpha,a} = \max\Big(\|A_{22}\|_{\alpha,a,D_{11}};\ \|A_{22}\|_{\alpha,a,D_{12}};\ \|A_{11}\|_{\alpha,a,D_{21}};$$
$$\|A_{11}\|_{\alpha,a,D_{22}};\ \|A_1\|^{\alpha,a};\ \|A_2\|^{\alpha,a};\ \|A\|^{1+\alpha,a}\Big)$$

<u>II.5 Lemma</u> $\mu_{\alpha,a}$ is a linear space.

<u>Proof</u> Obvious. We shall sometimes let $\mu_{\alpha,a} = \mu_a$.

<u>II.6 Lemma</u> If $A \in \mu_{\alpha,a}$, then

1) In D_{22}, $A(x,y) = \int_y^x A_{11}(t,y)(x-t)\,dt$. a.e.

2) In D_{21}, $A(x,y) = \int_{2a-y}^{x} A_{11}(t,y)(x-t)\,dt + A_1(2a-y,y)(x+y-2a)$
$+ A(2a-y,y)$ a.e.

3) In D_{12}, $A(x,y) = \int_y^{2a-x} A_{22}(x,t)(x-t)\,dt + A_2(x,2a-x)(x+y-2a)$
$+ A(x,2a-x)$ a.e.

4) In D_{11}, $A(x,y) = \int_y^x A_{22}(x,t)(t-y)\,dt$ a.e.

<u>Proof</u> Follows simply from the measurability and boundedness of the second derivative, the absolute continuity properties, the continuity in a neighborhood of $y = 2a-x$, and property II.4.4.

<u>II.7 Lemma</u> If $A \in \mu_{\alpha,a}$, then expression II.2 for $\Gamma(A)$ may be written as

1) In $D_{21} \cup D_{22}$

$$\Gamma(A) = \frac{A_1(x,y)}{(x-a)} - \nu \frac{A(x,y)}{(x-a)^2} - (x-a)^{\nu}(y-a)^{1-\nu} \int_a^y (\xi-a)^{\nu}\, d\xi$$

$$\left[\frac{A_{11}\left(a+\sqrt{(\xi-a)^2+(x-a)^2-(y-a)^2}\ ,\xi\right)}{\left(\sqrt{(\xi-a)^2+(x-a)^2-(y-a)^2}\right)^{\nu+2}} \quad \frac{-(2\nu+1)\ A_1\left(a+\sqrt{(\xi-a)^2+(x-a)^2-(y-a)^2}\ ,\xi\right)}{\left(\sqrt{(\xi-a)^2+(x-a)^2-(y-a)^2}\right)^{\nu+3}}\right.$$

$$\left. +\ \nu(\nu+2) \quad \frac{A\left(a+\sqrt{(\xi-a)^2+(x-a)^2-(y-a)^2}\ ,\xi\right)}{\left(\sqrt{(\xi-a)^2+(x-a)^2-(y-a)^2}\right)^{\nu+4}}\right] \quad \text{a.e.}$$

2) In $D_{11} \cup D_{12}$

$$\Gamma(A) = \frac{A_2(x,y)}{(y-a)} - (1-\nu)\ \frac{A(x,y)}{(y-a)^2} - (x-a)^{\nu}(y-a)^{1-\nu} \int_a^x (\eta-a)^{1-\nu}\ d\eta\ .$$

$$\left[\frac{A_{22}\left(\eta,a-\sqrt{(y-a)^2+(\eta-a)^2-(x-a)^2}\right)}{\left(-\sqrt{(y-a)^2+(\eta-a)^2-(x-a)^2}\right)^{3-\nu}} \quad \frac{-(3-2\nu)\ A_2\left(\eta,a-\sqrt{(y-a)^2+(\eta-a)^2-(x-a)^2}\right)}{\left(-\sqrt{(y-a)^2+(\eta-a)^2-(x-a)^2}\right)^{4-\nu}}\right.$$

$$\left. +(1-\nu)(3-\nu)\ \frac{A\left(\eta,a-\sqrt{(y-a)^2+(\eta-a)^2-(x-a)^2}\right)}{\left(-\sqrt{(y-a)^2+(\eta-a)^2-(x-a)^2}\right)^{5-\nu}}\right] \quad \text{a.e.}$$

<u>Proof</u> In D_{22}

$$\Gamma(A) = \frac{1}{(x-a)^{1-\nu}(y-a)^{\nu}}\ \frac{\partial^2}{\partial x\partial y}\left[\int_a^y \frac{d\xi\,(\xi-a)^{\nu}}{\left(+\sqrt{(\xi-a)^2+(x-a)^2-(y-a)^2}\right)^{\nu}}\right.$$

$$\left.\cdot\int_{\xi}^{a+\sqrt{(\xi-a)^2+(x-a)^2-(y-a)^2}} dt\ A_{11}(t,\xi)\ \left(a+\sqrt{(\xi-a)^2+(x-a)^2-(y-a)^2}\ -t\right)\right] \quad \text{a.e.}$$

If we carry out the differentiation we get expression 1) above. The integrals in 1) converge if $x > y$ (i.e. almost everywhere) by Fubini's theorem. Hence, by the dominated convergence theorem, the two expressions for $\Gamma(A)$ in D_{22} are equal (a.e.).

The same reasoning is valid in $D_{11} \cup D_{12} \cup D_{21}$.

<u>II.8 Theorem</u> Let $A \in \mu_{\alpha}$ for $\alpha > 2$. Then $\Gamma(A)$ is a Volterra kernel on $L_p[0,1]$ $\forall$ p, $1 \le p \le \infty$, and $\exists$ $C(\alpha,a,\nu) > 0$ such that

$$\|\Gamma(A)\|_{\alpha/2,\infty} \le C(\alpha,a,\nu)\ |A|^{\alpha,a} \quad \text{a.e.}$$

In what follows, all estimates will be valid a.e.

Proof $\Gamma(A)$ is a measurable function on D. If we can show $\|\Gamma(A)\|_{\alpha/2,\infty} < \infty$, it follows that $\Gamma(A)$ is a Volterra kernel on all $L_p[0,1]$ by Lemma I.7.

In D_{11}:

1) $|A(x,y)| \le \|A_{22}\|_{\alpha,a,D_{11}} \int_y^x (x-t)^{\alpha-1}(t-y)\,dt$

$$= \frac{(x-y)^{\alpha+1}}{\alpha(\alpha+1)} \|A_{22}\|_{\alpha,a,D_{11}}$$

2) $|A_2(x,y)| \le \|A_{22}\|_{\alpha,a,D_{11}} \int_y^x (x-t)^{\alpha-1}\,dt$

$$= \frac{(x-y)^{\alpha}}{\alpha} \|A_{22}\|_{\alpha,a,D_{11}}$$

3) $|A_{22}(x,y)| \le \|A_{22}\|_{\alpha,a,D_{11}} (x,y)^{\alpha-1}$

4) $0 < \frac{1}{a-y} \le \frac{1}{x-y}$ in D_{11} so

5) $$\left|\frac{A_2(x,y)}{(y-a)}\right| \le \frac{(x-y)^{\alpha-1}}{\alpha} \|A_{22}\|_{\alpha,a,D_{11}}$$

6) $$\left|\frac{(1-\nu)A(x,y)}{(y-a)^2}\right| \le \frac{(1-\nu)(x-y)^{\alpha-1}\|A_{22}\|_{\alpha,a,D_{11}}}{\alpha(\alpha+1)}$$

7) $$\left|\int_a^x (\eta-a)^{1-\nu}\,d\eta\, \frac{A_{22}\left(\eta, a-\sqrt{(y-a)^2+(\eta-a)^2-(x-a)^2}\right)}{\left(-\sqrt{(y-a)^2+(\eta-a)^2-(x-a)^2}\right)^{3-\nu}}\right| \le$$

$$\le \int_x^a (a-\eta)^{1-\nu} \frac{d\eta\left(\eta-a+\sqrt{(y-a)^2+(\eta-a)^2-(x-a)^2}\right)^{\alpha-1}}{\left(\sqrt{(y-a)^2+(\eta-a)^2-(x-a)^2}\right)^{3-\nu}} \|A_{22}\|_{\alpha,a,D_{11}}$$

(For this estimate, and for all estimates in $D_{11} \cup D_{22}$, we permit $\frac{1}{2} < \alpha \le 2$ in order that we may prove a subsequent theorem).

If $1 \le \alpha$

$$\le \|A_{22}\|_{\alpha,a,D_{11}} \left(\sqrt{(y-a)^2-(x-a)^2}\right)^{\alpha-1} \cdot \int_x^a \frac{(a-\eta)^{1-\nu}\, d\eta}{\left(\sqrt{(y-a)^2+(\eta-a)^2-(x-a)^2}\right)^{3-\nu}}$$

8) $$\int_x^a \frac{(a-\eta)^{1-\nu}\, d\eta}{\left(\sqrt{(y-a)^2+(\eta-a)^2-(x-a)^2}\right)^{3-\nu}}$$

$$= \frac{1}{2-\nu}\left[\left|\frac{(x-a)^{2-\nu}}{(y-a)^{3-\nu}}\right| + (3-\nu)\int_x^a \frac{(a-\eta)^{3-\nu}\, d\eta}{\left(\sqrt{(y-a)^2+(\eta-a)^2-(x-a)^2}\right)^{5-\nu}}\right]$$

$$\le \frac{1}{2-\nu}\left[\frac{(a-x)^{2-\nu}}{(a-y)^{3-\nu}} + (3-\nu)\int_x^a \frac{(a-\eta)\, d\eta}{\left(\sqrt{(y-a)^2+(\eta-a)^2-(x-a)^2}\right)^{3}}\right]$$

$$= \frac{1}{2-\nu}\left[\frac{(a-x)^{2-\nu}}{(a-y)^{3-\nu}} + (3-\nu)\left(\frac{1}{\sqrt{(y-a)^2-(x-a)^2}} - \frac{1}{a-y}\right)\right]$$

$$\le \frac{(3-\nu)}{(2-\nu)} \frac{1}{\sqrt{(y-a)^2-(x-a)^2}}$$

Call the three integral terms in II.7.2 term a, term b, and term c, respectively. Thus in D_{11} if $1 \le \alpha$

9) $$|\text{term a}| \le \frac{(a-x)^{\nu}(a-y)^{1-\nu}(x-y)^{\alpha/2\,-1}(2a)^{\alpha/2}}{2a-x-y} \frac{3-\nu}{2-\nu} \|A_{22}\|_{\alpha,a,D_{11}}$$

$$\le (x-y)^{\alpha/2\,-1}(2a)^{\alpha/2} \frac{3-\nu}{2-\nu} \|A_{22}\|_{\alpha,a,D_{11}}$$

Now if $\frac{1}{2} < \alpha \le 1$

7') $$\le (x-y)^{\alpha-1} \int_x^a \frac{(a-\eta)^{1-\nu} d\eta\, \|A_{22}\|_{\alpha,a,D_{11}}}{\left(\sqrt{(y-a)^2+(\eta-a)^2-(x-a)^2}\right)^{3-\nu}}$$

8') $$\le \frac{(x-y)^{\alpha-1}}{(2-\nu)}\left[\frac{(a-x)^{2-\nu}}{(a-y)^{3-\nu}} + (3-\nu)\int_x^a \frac{(a-\eta)^{3-\nu}\, d\eta}{\left(\sqrt{(y-a)^2+(\eta-a)^2-(x-a)^2}\right)^{5-\nu}}\right] \cdot \|A_{22}\|_{\alpha,a,D_{11}}$$

$$\le \frac{(x-y)^{\alpha-\,3/2}\, \frac{(3-\nu)}{(2-\nu)} \|A_{22}\|_{\alpha,a,D_{11}}}{\sqrt{2a-x-y}}$$

9') $$|\text{term } a| \le \sqrt{a}\,(x-y)^{\alpha-3/2}\,\frac{3-\nu}{2-\nu}\,\|A_{22}\|_{\alpha,a,D_{11}}$$

$$\text{if } \tfrac{1}{2} < \alpha \le 1$$

Next consider terms (b) and (c). It can easily be shown that

10) $$|\text{term } b| \le \frac{3-2\nu}{\alpha}\,|\text{term } a|$$

and

11) $$|\text{term } c| \le \frac{(1-\nu)(3-\nu)}{\alpha(\alpha+1)}\,|\text{term } a|$$

Thus, if $\alpha > 2$, we have

12) $$\|\Gamma(A)\|_{\alpha/2,a,D_{11}}$$

$$\le \|A_{22}\|_{\alpha,a,D_{11}}\, a^{\alpha/2}\left[1/\alpha + \frac{(1-\nu)}{\alpha(\alpha+1)} + 2^{\alpha/2}\,\frac{3-\nu}{2-\nu}\left(1+\frac{3-2\nu}{\alpha}+\frac{(1-\nu)(3-\nu)}{\alpha(\alpha+1)}\right)\right]$$

13) $$\|\Gamma(A)\|_{\alpha/2,a,D_{11}} \le \|A_{22}\|_{\alpha,a,D_{11}}\; C_{11}(\alpha,a,\nu)$$

Now we consider D_{12}

14) $$|A(x,y)| \le \|A_{22}\|_{\alpha,a,D_{12}}\int_y^{2a-x}(x-t)^{\alpha-1}(t-y)\,dt$$

$$+ \|A_2\|^{\alpha,a}(x-a)^{\alpha}(2a-x-y)$$

$$+ \|A\|^{1+\alpha,a}(x-a)^{1+\alpha}$$

$$\le \frac{(x-y)^{\alpha+1}}{\alpha(\alpha+1)}\,\|A_{22}\|_{\alpha,a,D_{12}} + \|A_2\|^{\alpha,a}(x-a)^{\alpha}(2a-x-y)$$

$$+ \|A\|^{1+\alpha,a}(x-a)^{1+\alpha}$$

15) $$|A_2(x,y)| \le \frac{(x-y)^{\alpha}}{\alpha}\,\|A_{22}\|_{\alpha,a,D_{12}} + \|A_2\|^{\alpha,a}(x-a)^{\alpha}$$

16) $$|A_{22}(x,y)| \le (x-y)^{\alpha-1}\,\|A_{22}\|_{\alpha,a,D_{12}}$$

17) $$0 \le 2(x-a) \le (x-y) \le 2(a-y)$$

18) $$\left|\frac{A_2(x,y)}{(y-a)}\right| \le 2/\alpha\ (x-y)^{\alpha-1}\ \|A_{22}\|_{\alpha,a,D_{12}} + (\tfrac{1}{2})^{\alpha-1}(x-y)^{\alpha-1}\ \|A_2\|^{\alpha,a}$$

19) $$(1-\nu)\ \left|\frac{A(x,y)}{(y-a)^2}\right| \le \left(\frac{4(x-y)^{\alpha-1}}{\alpha(\alpha+1)}\ \|A_{22}\|_{\alpha,a,D_{12}} + (\tfrac{1}{2})^{\alpha-1}(x-y)^{\alpha-1}\ \|A_2\|^{\alpha,a} + (\tfrac{1}{2})^{\alpha-1}(x-y)^{\alpha-1}\|A\|^{1+\alpha,a}\right)(1-\nu)$$

20) Consider term a

$$\left|\int_a^x (\eta-a)^{1-\nu}\ d\eta\ \frac{A_{22}\left(\eta, a-\sqrt{(y-a)^2+(\eta-a)^2-(x-a)^2}\right)}{\left(-\sqrt{(y-a)^2+(\eta-a)^2-(x-a)^2}\right)^{3-\nu}}\right|$$

$$\le \|A_{22}\|_{\alpha,a,D_{12}}\int_a^x (\eta-a)^{1-\nu}\ \frac{\left(\eta-a+\sqrt{(y-a)^2+(\eta-a)^2-(x-a)^2}\right)^{\alpha-1}}{\left(\sqrt{(y-a)^2+(\eta-a)^2-(x-a)^2}\right)^{3-\nu}}\ d\eta\ .$$

$$\le 2^{\alpha-1}\ \|A_{22}\|_{\alpha,a,D_{12}}\int_a^x \left(\sqrt{(y-a)^2+(\eta-a)^2-(x-a)^2}\right)^{\alpha-3}\ d\eta\ .$$

If $\alpha < 3$, $\le 2^{\alpha-1}\int_a^x \frac{d\eta}{(\eta-a)^{3-\alpha}} \le \frac{2^{\alpha-1}}{(\alpha-2)}\ (x-a)^{\alpha-2}\ \|A_{22}\|_{\alpha,a,D_{12}}$.

If $\alpha \ge 3$, $\le 2^{\alpha-1}(x-y)^{\alpha-3}(x-a)$

21) $$|\text{term a}|_{D_{12}} \le \max\left[2^{\alpha-2},\ \frac{2}{\alpha-2}\right](x-y)^{\alpha-1}\ \|A_{22}\|_{\alpha,a,D_{12}}\ .$$

It can easily be shown that

22) $$|\text{term b}|_{D_{12}} \le (3-2\nu)\left(2/\alpha\ |\text{term a}|_{D_{12}} + (\tfrac{1}{2})^{\alpha-2}\ \frac{\|A_2\|^{\alpha,a}}{(\alpha-2)}\ (x-y)^{\alpha-1}\right)$$

$|f|_{D_{ij}}$ merely indicates the absolute value of $f(x,y)$ at some point in D_{ij}.

We also have

$$23)\qquad |\text{term } c|_{D_{12}} \le (1-\nu)(3-\nu)\left[4\,\frac{|\text{term } a|_{D_{12}}}{\alpha(\alpha+1)} + (\tfrac{1}{2})^{\alpha-2}\,\frac{(x-y)^{\alpha-1}}{\alpha-2}\left(\|A_2\|^{\alpha,a} + \|A\|^{1+\alpha,a}\right)\right]$$

Finally:

$$24)\qquad \|\Gamma(A)\|_{\alpha/2,a,D_{12}} \le \min[2a,1]^{\alpha/2}\left[\|A_{22}\|_{\alpha,a,D_{12}}\left(2/\alpha + \frac{4}{\alpha(\alpha+1)} + \max[2^{\alpha-2},2/\alpha-2]\left(1+ 2/\alpha(3-2\nu)+\frac{4(1-\nu)(3-\nu)}{\alpha(\alpha+1)}\right)\right) + \|A_2\|^{\alpha,a}\left((\tfrac{1}{2})^{\alpha-1}\left[(2-\nu)+\frac{6-6\nu+\nu^2}{\alpha-2}\right]\right) + \|A\|^{1+\alpha,a}(\tfrac{1}{2})^{\alpha-2}\left(1-\nu+\frac{(1-\nu)(3-\nu)}{\alpha-2}\right)\right]$$

$$25)\qquad \|\Gamma(A)\|_{\alpha/2,a,D_{12}} \le C_{12}^{(1)}(\alpha,a,\nu)\,\|A_{22}\|_{\alpha,a,D_{12}} + C_{12}^{(2)}(\alpha,a,\nu)\,\|A_2\|^{\alpha,a} + C_{12}^{(3)}(\alpha,a,\nu)\,\|A\|^{1+\alpha,a}$$

The estimates for D_{21} and D_{22} are completely analogous to those for D_{12} and D_{11}. The results are

$$26)\qquad \|\Gamma(A)\|_{\alpha/2,a,D_{21}} \le C_{12}^{(1)}(\alpha,1-a,1-\nu)\|A_{11}\|_{\alpha,a,D_{21}} + C_{12}^{(2)}(\alpha,1-a,1-\nu)\|A_1\|^{\alpha,a} + C_{12}^{(3)}(\alpha,1-a,1-\nu)\|A\|^{1+\alpha,a}$$

27) If $\alpha > 1$, $\|\Gamma(A)\|_{\alpha/2,a,D_{22}} \le C_{11}(\alpha,1-a,1-\nu)\|A_{11}\|_{\alpha,a,D_{22}}$

28) $$\|\Gamma(A)\|_{\alpha/2,\infty} = \sup_{\substack{i=1,2\\ j=1,2}} (\|\Gamma(A)\|_{\alpha/2,a,D_{ij}})$$

$$\leq \max(C_{11}(\alpha,a,\nu)\|A_{22}\|_{\alpha,a,D_{11}}, C_{11}(\alpha,1-a,1-\nu)\|A_{11}\|_{\alpha,a,D_{22}},$$

$$C_{12}^{(1)}(\alpha,a,\nu)\|A_{22}\|_{\alpha,a,D_{12}}, C_{12}^{(2)}(\alpha,a,\nu)\|A_2\|^{\alpha,a},$$

$$C_{12}^{(3)}(\alpha,a,\nu)\|A\|^{1+\alpha,a}, C_{12}^{(1)}(\alpha,1-a,1-\nu)\|A_{11}\|_{\alpha,a,D_{21}},$$

$$C_{12}^{(2)}(\alpha,1-a,1-\nu)\|A_1\|^{\alpha,a}, C_{12}^{(3)}(\alpha,1-a,1-\nu)\|A\|^{1+\alpha,a})$$

$$\leq C(\alpha,a,\nu)\ |A|^{\alpha,a}$$

<u>II.9 Lemma</u> Let $A \in \mu_\alpha$ for $\frac{1}{2} < \alpha < \infty$. Then $\Gamma(A)$ is measurable on $D_{11} \cup D_{22}$ and $\exists$ a constant $d_{11}(\alpha,a,\nu)$ such that

$$\|\Gamma(A)\|_{\min(\alpha/2,\alpha-\frac{1}{2}),a,D_{11}} \leq d_{11}(\alpha,a,\nu)\|A_{22}\|_{\alpha,a,D_{11}}$$

and

$$\|\Gamma(A)\|_{\min(\alpha/2,\alpha-\frac{1}{2}),a,D_{22}} \leq d_{11}(\alpha,1-a,1-\nu)\|A_{11}\|_{\alpha,a,D_{22}}$$

<u>Proof</u> Measurability is clear. Estimates 5 and 6 of the last lemma are valid if $\alpha > 0$. Estimates 7' to 9' and 10 and 11 are valid if $\alpha > \frac{1}{2}$.

$$d_{11}(\alpha,a,\nu) = C_{11}(\alpha,a,\nu) \quad \text{if } \alpha > 1$$

if $\frac{1}{2} < \alpha \leq 1$, $d_{11}(\alpha,a,\nu) = \sqrt{a}\left[\frac{3-\nu}{2-\nu}\left(1+\frac{3-2\nu}{\alpha}+\frac{(1-\nu)(3-\nu)}{\alpha(\alpha+1)}\right)+1/\alpha+\frac{(1-\nu)}{\alpha(\alpha+1)}\right]$

Again, it may be shown that in D_{22}, $d_{22}(\alpha,a,\nu) = d_{11}(\alpha,1-a,1-\nu)$ if $\frac{1}{2} < \alpha \leq 1$.

<u>II.10 Definition</u> If $A \in \mu_{\alpha,a}$ for $\alpha > \frac{1}{2}$ and either

(1) $a = 0$ or $a = 1$, or

(2) $0 < a < 1$ and $A \equiv 0$ in $D_{21}^{(a)} \cup D_{12}^{(a)}$ a.e.

then we say $A \in L_{\alpha,a}$.

II.11 Theorem If $A \in L_{\alpha,a}$ for $\alpha > \frac{1}{2}$, then $\Gamma(A)$ is the kernel of a Volterra integral operator on all $L_p[0,1]$, $1 \le p \le \infty$ and

$$\|\Gamma(A)\|_{\min(\alpha-\frac{1}{2},\alpha/2),a,D_{11}} \le d_{11}(\alpha,a,\nu)\|A_{22}\|_{\alpha,a,D_{11}}$$

$$\|\Gamma(A)\|_{\min(\alpha-\frac{1}{2},\alpha/2),a,D_{22}} \le d_{11}(\alpha,1-a,1-\nu)\|A_{11}\|_{\alpha,a,D_{22}}.$$

Proof By Lemma II.3, $\Gamma_{21}(A) = \Gamma_{12}(A) = 0$ a.e. If $a = 0$, then $\Gamma(A) = \Gamma_{22}(A)$, if $a = 1$, then $\Gamma(A) = \Gamma_{11}(A)$. We then use Lemma II.9 and the result follows immediately.

II.12 Theorem If $A \in \mu_\alpha$ for $\alpha > 2$, or if $A \in L_\alpha$ for $\alpha > \frac{1}{2}$, then $\Gamma(A)$ solves the commutator equation a.e. on D

$$\int_y^x [(x-a)^\nu (t-a)^{1-\nu}\Gamma(t,y) - \Gamma(x,t)(t-a)^\nu (y-a)^{1-\nu}]dt = A(x,y)$$

a.e. on D

Proof (All the following equalities are valid a.e.)

In D_{11}

1) $$Q*\Gamma(A) = \int_y^x (x-a)^\nu (t-a)^{1-\nu}\, dt \frac{\partial}{\partial t} \int_a^t (\eta-a)^{1-\nu}\, d\eta$$

$$\left[\frac{A_2\left(\eta, a-\sqrt{(y-a)^2+(\eta-a)^2-(t-a)^2}\right)}{\left(-\sqrt{(y-a)^2+(\eta-a)^2-(t-a)^2}\right)^{2-\nu}} - (1-\nu)\frac{A\left(\eta, a-\sqrt{(y-a)^2+(\eta-a)^2-(t-a)^2}\right)}{\left(-\sqrt{(y-a)^2+(\eta-a)^2-(t-a)^2}\right)^{3-\nu}}\right]$$

$$= (x-a)^\nu (y-a)^{1-\nu} \int_a^x (\eta-a)^{1-\nu} d\eta \left[\frac{A_2\left(\eta, a-\sqrt{(y-a)^2+(\eta-a)^2-(x-a)^2}\right)}{\left(-\sqrt{(y-a)^2+(\eta-a)^2-(x-a)^2}\right)^{2-\nu}}\right.$$

$$\left. - (1-\nu)\frac{A\left(\eta, a-\sqrt{(y-a)^2+(\eta-a)^2-(x-a)^2}\right)}{\left(-\sqrt{(y-a)^2+(\eta-a)^2-(x-a)^2}\right)^{3-\nu}}\right]$$

because $\lim_{y\to\eta} A_2(\eta,y) = \lim_{y\to\eta} A(\eta,y) = 0$ if $\eta < a$

2) $$-\Gamma(A)*Q = -\int_y^x \frac{(y-a)^{1-\nu}}{(x-a)^{1-\nu}}\,dt\,\frac{\partial}{\partial t}\left[\frac{A(x,t)(x-a)^{1-\nu}}{(t-a)^{1-\nu}} - (x-a)\int_a^x (\eta-a)^{1-\nu}\,d\eta\right.$$

$$\left.\left[\frac{A_2\left(\eta, a-\sqrt{(t-a)^2+(\eta-a)^2-(x-a)^2}\right)}{\left(-\sqrt{(t-a)^2+(\eta-a)^2-(x-a)^2}\right)^{2-\nu}} - (1-\nu)\frac{A\left(\eta, a-\sqrt{(t-a)^2+(\eta-a)^2-(x-a)^2}\right)}{\left(-\sqrt{(t-a)^2+(\eta-a)^2-(x-a)^2}\right)^{3-\nu}}\right]\right]$$

$$= A(x,y) - (x-a)^{\nu}(y-a)^{1-\nu}\int_a^x (\eta-a)^{1-\nu}\,d\eta$$

$$\left[\frac{A_2\left(\eta, a-\sqrt{(y-a)^2+(\eta-a)^2-(x-a)^2}\right)}{\left(-\sqrt{(y-a)^2+(\eta-a)^2-(x-a)^2}\right)^{2-\nu}} - (1-\nu)\frac{A\left(\eta, a-\sqrt{(y-a)^2+(\eta-a)^2-(x-a)^2}\right)}{\left(-\sqrt{(y-a)^2+(\eta-a)^2-(x-a)^2}\right)^{3-\nu}}\right]$$

3) $Q*\Gamma(A) - \Gamma(A)*Q = A$ a.e. in D_{11} where we write Q for the kernel of Q.

Next, consider D_{12}

4) $$Q*\Gamma(A) = \int_y^x (x-a)^{\nu}(y-a)^{1-\nu}\,dt\,\frac{\partial}{\partial t}\int_a^t (\eta-a)^{1-\nu}\,d\eta$$

$$\left[\frac{A_2\left(\eta, a-\sqrt{(y-a)^2+(\eta-a)^2-(t-a)^2}\right)}{\left(-\sqrt{(y-a)^2+(\eta-a)^2-(t-a)^2}\right)^{2-\nu}} - (1-\nu)\frac{A\left(\eta, a-\sqrt{(y-a)^2+(\eta-a)^2-(t-a)^2}\right)}{\left(-\sqrt{(y-a)^2+(\eta-a)^2-(t-a)^2}\right)^{3-\nu}}\right]$$

$$= (x-a)^{\nu}(y-a)^{1-\nu}\int_a^x (\eta-a)^{1-\nu}\left[\frac{A_2\left(\eta, a-\sqrt{(y-a)^2+(\eta-a)^2-(x-a)^2}\right)}{\left(-\sqrt{(y-a)^2+(\eta-a)^2-(x-a)^2}\right)^{2-\nu}}\right.$$

$$\left. - (1-\nu)\frac{A\left(\eta, a-\sqrt{(y-a)^2+(\eta-a)^2-(x-a)^2}\right)}{\left(-\sqrt{(y-a)^2+(\eta-a)^2-(x-a)^2}\right)^{3-\nu}}\right]d\eta$$

5) $$-\Gamma(A)*Q = -\int_y^{2a-x} \frac{(y-a)^{1-\nu}}{(x-a)^{1-\nu}}\,dt\,\frac{\partial^2}{\partial x\partial t}\int_a^x (\eta-a)^{1-\nu}$$

$$\frac{A\left(\eta, a-\sqrt{(t-a)^2+(\eta-a)^2-(x-a)^2}\right)}{\left(-\sqrt{(t-a)^2+(\eta-a)^2-(x-a)^2}\right)^{1-\nu}}\,d\eta - \int_{2a-x}^x \frac{(y-a)^{1-\nu}}{(x-a)^{1-\nu}}\,dt$$

$$\left[\frac{\partial^2}{\partial x\partial t}\int_a^t \frac{(\xi-a)^\nu A\left(a+\sqrt{(\xi-a)^2+(x-a)^2-(t-a)^2}\;,\xi\right)}{\left(\sqrt{(\xi-a)^2+(x-a)^2-(t-a)^2}\right)^\nu}\,d\xi\right]$$

Let the two integral terms above be terms g and h respectively.

6) $$\text{term g} = -\int_y^{2a-x}\frac{(y-a)^{1-\nu}}{(x-a)^{1-\nu}}\,dt\,\frac{\partial}{\partial t}\left[\frac{(x-a)^{1-\nu}}{(t-a)^{1-\nu}}A(x,t)-(x-a)\int_a^x(\eta-a)^{1-\nu}\,d\eta\right.$$

$$\left.\left[\frac{A_2\left(\eta,a-\sqrt{(t-a)^2+(\eta-a)^2-(x-a)^2}\right)}{\left(-\sqrt{(t-a)^2+(\eta-a)^2-(x-a)^2}\right)^{2-\nu}}-(1-\nu)\frac{A\left(\eta,a-\sqrt{(t-a)^2+(\eta-a)^2-(x-a)^2}\right)}{\left(-\sqrt{(t-a)^2+(\eta-a)^2-(x-a)^2}\right)^{3-\nu}}\right]\right]$$

$$= A(x,y)-(x-a)^\nu(y-a)^{1-\nu}\int_a^x(\eta-a)^{1-\nu}\,d\eta$$

$$\left[\frac{A_2\left(\eta,a-\sqrt{(y-a)^2+(\eta-a)^2-(x-a)^2}\right)}{\left(-\sqrt{(y-a)^2+(\eta-a)^2-(x-a)^2}\right)^{2-\nu}}-(1-\nu)\frac{A\left(\eta,a-\sqrt{(y-a)^2+(\eta-a)^2-(x-a)^2}\right)}{\left(-\sqrt{(y-a)^2+(\eta-a)^2-(x-a)^2}\right)^{3-\nu}}\right]$$

$$-\frac{(y-a)^{1-\nu}}{(a-x)^{1-\nu}}A(x,2a-x)+(x-a)^\nu(y-a)^{1-\nu}\int_a^x(\eta-a)^{1-\nu}d\eta\left[\frac{A_2(\eta,2a-\eta)}{(a-\eta)^{2-\nu}}-(1-\nu)\frac{A(\eta,2a-\eta)}{(a-\eta)^{3-\nu}}\right]$$

7) $$\text{term h} = -\int_{2a-x}^x(y-a)^{1-\nu}(x-a)^\nu\frac{\partial}{\partial t}\int_a^t(\xi-a)^\nu$$

$$\left[\frac{A_1\left(a+\sqrt{(\xi-a)^2+(x-a)^2-(t-a)^2}\;,\xi\right)}{\left(\sqrt{(\xi-a)^2+(x-a)^2-(t-a)^2}\right)^{1+\nu}}-\nu\frac{A\left(a+\sqrt{(\xi-a)^2+(x-a)^2-(t-a)^2}\;,\xi\right)}{\left(\sqrt{(\xi-a)^2+(x-a)^2-(t-a)^2}\right)^{2+\nu}}\right]d\xi$$

$$=(y-a)^{1-\nu}(x-a)^\nu\int_a^{2a-x}(\xi-a)^\nu\left[\frac{A_1(2a-\xi,\xi)}{(a-\xi)^{1+\nu}}-\nu\frac{A(2a-\xi,\xi)}{(a-\xi)^{\nu+2}}\right]d\xi$$

we wish to show

8) $$0 = -\frac{(y-a)^{1-\nu}}{(a-x)^{1-\nu}}A(x,2a-x)+(x-a)^\nu(y-a)^{1-\nu}$$

$$\left(\int_a^{2a-x} (\xi-a)^\nu \left[\frac{A_1(2a-\xi,\xi)}{(a-\xi)^{1+\nu}} - \nu \frac{A(2a-\xi,\xi)}{(a-\xi)^{\nu+2}}\right] d\xi \right.$$

$$\left. + \int_a^x (\eta-a)^{1-\nu} \left[\frac{A_2(\eta,2a-\eta)}{(a-\eta)^{2-\nu}} - (1-\nu) \frac{A(\eta,2a-\eta)}{(a-\eta)^{3-\nu}}\right] d\eta \right)$$

Call the five terms in 8), i, j, k, ℓ, m, respectively. Either $0 \le \arg f < 2\pi$, or $-2\pi < \arg f \le 0$ throughout. We treat both cases simultaneously

$$\text{term i} = -\frac{(y-a)^{1-\nu}}{(x-a)^{1-\nu}} \frac{A(x,2a-x)}{e^{\pm i\pi(1-\nu)}} = e^{\pm i\pi\nu} \frac{(y-a)^{1-\nu}}{(x-a)^{1-\nu}} A(x,2a-x)$$

$$\text{term j} = e^{\pm i\pi\nu}(x-a)^\nu(y-a)^{1-\nu}\int_a^{2a-x} \frac{A_1(2a-\xi,\xi)}{(a-\xi)} d\xi$$

$$= -(x-a)^\nu(y-a)^{1-\nu} e^{\pm i\pi\nu} \int_a^x \frac{A_1(\eta,2a-\eta)}{(\eta-a)} d\eta$$

$$\text{term k} = -\nu e^{\pm i\pi\nu}(x-a)^\nu(y-a)^{1-\nu}\int_a^{2a-x} \frac{A(2a-\xi,\xi)}{(\xi-a)^2} d\xi$$

$$= \nu e^{\pm i\pi\nu}(x-a)^\nu(y-a)^{1-\nu} \int_a^x \frac{A(\eta,2a-\eta)}{(\eta-a)^2} d\eta$$

$$\text{term } \ell = e^{\mp i\pi(2-\nu)}(x-a)^\nu(y-a)^{1-\nu} \int_a^x \frac{A_2(\eta,2a-\eta)}{(\eta-a)} d\eta$$

$$= e^{\pm i\pi\nu}(x-a)^\nu(y-a)^{1-\nu} \int_a^x \frac{A_2(\eta,2a-\eta)}{(\eta-a)} d\eta$$

$$\text{term m} = (1-\nu) e^{\pm i\pi\nu}(x-a)^\nu(y-a)^{1-\nu} \int_a^x \frac{A(\eta,2a-\eta)}{(\eta-a)^2} d\eta$$

$$\text{total} = e^{\pm i\pi\nu}(x-a)^\nu(y-a)^{1-\nu} \left[\frac{A(x,2a-x)}{(x-a)} + \right.$$

$$+ \int_a^x \left[- \frac{A_1(\eta, 2a-\eta)}{(\eta-a)} + \frac{A(\eta,2a-\eta)}{(\eta-a)^2} + \frac{A_2(\eta,2a-\eta)}{(\eta-a)} \right] d\eta \Bigg]$$

but $$\int_a^x \frac{A(\eta,2a-\eta)}{(\eta-a)^2} d\eta = - \frac{A(x,2a-x)}{(x-a)} + \int_a^x \frac{d/d\eta\, A(\eta,2a-\eta)}{(\eta-a)} d\eta .$$

total = 0

or $Q*\Gamma(A) - \Gamma(A)*Q = A$ a.e. in D_{12}.

The proofs in D_{21} and D_{22} are analogous to those in D_{12} and D_{11} respectively.

III. SIMILARITY OF $KQ + T_B$ TO KQ

In this section we shall consider various classes of B's having the property that $A_B \in \mu_\alpha$ for $\alpha > 2$ or $A_B \in L_\alpha$ for $\alpha > \frac{1}{2}$. A_B satisfies

$$A_B + 1/K\, B*\Gamma(A_B) = B$$

<u>III.1 Definition</u> If $A \in \mu_\alpha$ for $\alpha > 2$, or $A \in L_\alpha$ for $\alpha > \frac{1}{2}$, and $\|B\|_{\gamma,\infty} < \infty$ for some $\gamma > 0$, we define

$$\tilde{T}_B(A) = B*\Gamma(A)$$

<u>III.2 Lemma</u> $\tilde{T}_B(A)$ is a bilinear function of A and B. Furthermore, it is the kernel of a Volterra integral operator on all $L_p[0,1]$ and

$$\|\tilde{T}_B(A)\|_{\min(\alpha-\frac{1}{2},\alpha/2)+\gamma,\infty} \le \beta(\min(\alpha-\tfrac{1}{2},\alpha/2),\gamma)\cdot\|B\|_{\gamma,\infty}\|\Gamma(A)\|_{\min(\alpha-\frac{1}{2},\alpha/2),\infty}$$

<u>Proof</u> Bilinearity is trivial. The second statement is a consequence of I.6, I.7, II.8 and II.11.

<u>III.3 Definition</u> For any a satisfying $0 \le a \le 1$ and ϵ with $0 < \epsilon < 1$, let

n_1 = smallest integer $\ge \max\left(a^2/\epsilon^2 - 1, 0\right)$

n_2 = smallest integer $\ge \max\left(\min \frac{(2a,1)}{\epsilon} - 1, 0\right)$

$n_0 = \max[n_1 + n_2 - 1, 0]$

III.4 Definition Let $R_{a,\epsilon}$ consist of functions $B(x,y)$ defined on D with the following properties

1) $B(x,y) \equiv 0$ if $(x-y) \le \epsilon$ a.e.

2) $\frac{\partial^j}{\partial y^j} B(x,y)$ is absolutely continuous w.r. to y a.e. in $D_{12} \cup D_{11}$, for $j = 0,1, \quad , 2n_0+1$

3) $\frac{\partial^k}{\partial x^k} B(x,y)$ is absolutely continuous w.r. to x a.e. in $D_{21} \cup D_{22}$, for $k = 0,1$.

4) $B(x,y)$, $B_1(x,y)$, and $\frac{\partial^j B}{\partial y^j}(x,y)$ for $j = 1,2,\ldots,2n_0+1$ are all continuous functions of (x,y) in the closure of some neighborhood of the line $y = 2a-x$ contained in D. (If $a = 0$ or $a = 1$, this requirement is omitted).

5) $\frac{\partial^{2n_0+2}}{\partial y^{2n_0+2}} B(x,y)$ is measurable on $D_{11} \cup D_{12}$

$\frac{\partial^2}{\partial x^2} B(x,y)$ is measurable on $D_{21} \cup D_{22}$

$$\operatorname*{ess\,sup}_{D_{11} \cup D_{12}} \left| \frac{\partial^{2n_0+2}}{\partial y^{2n_0+2}} B(x,y) \right| < \infty$$

$$\operatorname*{ess\,sup}_{D_{21} \cup D_{22}} \left| \frac{\partial^2}{\partial x^2} B(x,y) \right| < \infty$$

$R_{a,\epsilon}$ is obviously a linear space.

III.5 Lemma If $B \in R_{a,\epsilon}$, then $B, \tilde{T}_B(B), \ldots, \tilde{T}_B^{n_0}(B)$ all belong to $\mu_{\alpha,a}$ for any $\alpha > 2$.

Proof We shall show $\forall\, n \ni 0 \le n \le n_0$

1) $\tilde{T}_B^{\,n}(B) \equiv 0$ if $(x-y) \le \epsilon$ a.e.

2) $\frac{\partial^j}{\partial y^j} (\tilde{T}_B^{\,n}(B))(x,y)$ is absolutely continuous w.r. to y in $D_{11} \cup D_{12}$ a.e., for $j = 0,1,\ldots,2(n_0-n)+1$

3) $\frac{\partial^k}{\partial x^k} (\tilde{T}_B^{\,n}(B))(x,y)$ is absolutely continuous w.r. to x in $D_{21} \cup D_{22}$ a.e., for $k = 0,1$.

4) $\tilde{T}_B^{\,n}(B)(x,y)$, $\frac{\partial}{\partial x}(\tilde{T}_B^{\,n}(B)(x,y))$, $\frac{\partial^j}{\partial y^j}(\tilde{T}_B^{\,n}(B)(x,y))$

$j = 1,2,\ldots,2(n_0-n)+1$ are all continuous in the closure of a neighborhood of $y = 2a-x$ in D (if $a \neq 0$ and $a \neq 1$).

5)
$$\operatorname*{ess\,sup}_{D_{11}\cup D_{12}} \left| \frac{\partial^{2(n_0-n)+2}}{\partial y^{2(n_0-n)+2}} \tilde{T}_B^{\,n}(B)(x,y) \right| < \infty$$

$$\operatorname*{ess\,sup}_{D_{21}\cup D_{22}} \left| \frac{\partial^2}{\partial x^2} \tilde{T}_B^{\,n}(B)(x,y) \right| < \infty$$

$\frac{\partial^{2(n_0-n)+2}}{\partial y^{2(n_0-n)+2}} \tilde{T}_B^{\,n}(B)(x,y)$ is measurable on $D_{11} \cup D_{12}$

$\frac{\partial^2}{\partial x^2} \tilde{T}_B^{\,n}(B)(x,y)$ is measurable on $D_{21} \cup D_{22}$

Suppose we have shown 1) to 5) to be valid. Keep in mind that for $\alpha \geq 1$

6)
$$\operatorname*{ess\,sup}_{D_{ij}} |A(x,y)(x-y)^{1-\alpha}| \leq \epsilon^{1-\alpha} \operatorname*{ess\,sup}_{D_{ij}} |A(x,y)|$$

for A in $R_{a,\epsilon}$. It is then simple to show that $|\tilde{T}_B^{\,n}(B)|^{\alpha,a} < \infty \quad \forall\, \alpha \geq 1$ and in fact that $\tilde{T}_B^{\,n}(B) \in \mu_{\alpha,a} \quad \forall\, \alpha > 2$.

Thus we need only prove that properties 1) to 5) hold for any B in $R_{a,\epsilon}$. We do this by finite induction.

If $n = 0$, $\tilde{T}_B^{\,0}(B) = B$ and 1) to 5) are the same as 1) to 5) of III.4. If $n_0 = 0$, we are finished. If not: Suppose we have shown properties 1 to 5 to be true for $j = 0,1,\ldots,n \le n_0-1$ i.e. $\tilde{T}_B^{\,n}(B) = A^{(n)}$ satisfies 1 to 5 for all $n \le n_0-1$. Then $A^{(n)} \in \mu_\alpha$, so

$$A^{(n+1)} = \int_y^x B(x,y)\Gamma(A^{(n)})(t,y)dt = \tilde{T}_B(A^{(n)})$$

exists.

7) $\quad A^{(n+1)}(x,y) \equiv 0$ if $(x-y) \le \epsilon$ obviously

8) In D_{22},
$$\begin{aligned} A^{(n+1)}(x,y) &= \int_y^x B(x,t)\Gamma(A^{(n)})(t,y)dt \\ &= \int_y^x dt \int_t^x B_1(s,t)\Gamma(A^{(n)})(t,y)ds \\ &= \int_y^x ds \int_y^s B_1(s,t)\Gamma(A^{(n)})(t,y)dt \quad \text{a.e.} \\ &= \int_y^x ds \frac{\partial}{\partial s} A^{(n)}(s,y) \quad \text{a.e.} \end{aligned}$$

Thus $A^{(n+1)}(x,y)$ is absolutely continuous w.r. to x in D_{22} a.e. The same thing works for $\dfrac{\partial A^{(n+1)}}{\partial x}(x,y)$. In D_{21}, the proof is slightly more complicated because it involves boundary terms. However the main idea is again a simple change of order of integration and the same results follow.

Let $\Gamma(A^{(n)}) = \Gamma^{(n)}$ in what follows.

9) In D_{21}

$$\begin{aligned} A_{11}^{(n+1)}(x,y) &= \int_y^a B_{11}(x,t)\Gamma_{(11)}^{(n)}(t,y)dt + \int_a^{2a-y} B_{11}(x,t)\Gamma_{(12)}^{(n)}(t,y)dt \\ &\quad + \int_{2a-y}^x B_{11}(x,t)\Gamma_{(21)}^{(n)}(t,y)dt \end{aligned}$$

$A_{11}^{(n+1)}(x,y)$ is obviously measurable. Boundedness follows from the induction hypothesis if we note that the apparent singularities caused by vanishing denominators in the $\Gamma_{(ij)}^{(n)}(t,y)$ are nullified because of property 1.

The same sort of thing works in D_{22} with

10) $$A_{11}(x,y) = \int_y^x B_{11}(x,t)\Gamma^{(n)}_{(22)}(t,y)dt$$

In D_{21}, if $x+y-2a \le \epsilon$, then

11) $$A^{(n+1)}(x,y) = \int_y^a B(x,t)\Gamma^{(n)}_{(11)}(t,y)dt + \int_a^{2a-y} B(x,t)\Gamma^{(n)}_{(12)}(t,y)dt$$

because $B(x,t) \equiv 0$ if $x-t \le x+y-2a \le \epsilon$

thus $$\frac{\partial^j A^{(n+1)}}{\partial y^j}(x,y) = \int_y^a B(x,t)\left(\frac{\partial}{\partial y}\right)^j \Gamma^{(n)}_{(11)}(t,y)dt + \int_a^{2a-y} B(x,t)\left(\frac{\partial}{\partial y}\right)^j \Gamma^{(n)}_{(12)}(t,y)dt$$

for $j = 0,1,\ldots,2(n_0-n)-1$.

The derivatives involved are y derivatives of $A^{(n)}(t,y)$ and

$$A^{(n)}\left(\eta, a-\sqrt{(y-a)^2+(\eta-a)^2-(t-a)^2}\right)$$

up to order $j+2 \le 2(n_0-n)+1$ which are absolutely continuous y by the induction hypothesis. Thus we have continuity (x,y) if $x+y-2a \le \epsilon$ in $\overline{D}_{21}$.

12) In D_{12},

$$A^{(n+1)}(x,y) = \int_y^a B(x,t)\Gamma^{(n)}_{(11)}(t,y)dt + \int_a^x B(x,t)\Gamma^{(n)}_{(12)}(t,y)dt$$

In $\overline{D}_{12}$ let us concern ourselves with the intersection of the closures of the two neighborhoods in which $B(x,t)$ and $B_1(x,t)$ are continuous (x,t). Now the same argument follows.

Hence $\frac{\partial^j}{\partial y^j} A^{(n+1)}(x,y)$ is continuous (x,y) in the closure of some neighborhood of $y = 2a-x$, for $0 \le j \le 2(n_0-n)-1$. We may use expressions 11) and 12) in the same manner to show $\frac{\partial}{\partial x} A^{(n+1)}(x,y)$ is continuous in the closure of some neighborhood of $y = 2a-x$. Now in D_{11}:

13) $$A^{(n+1)}(x,y) = \int_y^x B(x,t)\Gamma^{(n)}_{(11)}(t,y)dt$$

$$\frac{\partial^j A^{(n+1)}}{\partial y^j}(x,y) = \int_y^x B(x,t)\left(\frac{\partial}{\partial y}\right)^j \Gamma^{(n)}_{(11)}(t,y)dt$$

for $j = 0,1,\dots,2(n_0-n)$. Thus absolute continuity, boundedness, and measurability can be shown in a straightforward manner here and in D_{12}.

q.e.d.

III.6 Lemma $\widetilde{T}_B^{\,n_0+1}(B)$ exists as a Volterra kernel on all $L_p[0,1]$ such that $\|\widetilde{T}_B^{\,n_0+1}(B)\|_{\alpha,\infty} < \infty$ for any $\alpha > 0$ and $\widetilde{T}_B^{\,n_0+1}(B) \equiv 0$ in $D_{11} \cup D_{12}$.

Proof $\widetilde{T}_B^{\,n_0}(B) \in \mu_\alpha$ for any $\alpha > 0$. Hence everything follows except $\widetilde{T}_B^{\,n_0+1}(B) \equiv 0$ in $D_{11} \cup D_{12}$. This we shall prove by induction.

In $D_{11} \cup D_{12}$, we have $\widetilde{T}_B^{\,n_0+1}(B) \equiv 0$ if $(x-y) \le \epsilon$. If $\epsilon > \min(2a,1)$ then $n_0 = 0 = n_1 + n_2$ and $B \equiv 0$ in $D_{11} \cup D_{12}$, hence $\Gamma(B) = 0$ in $D_{11} \cup D_{12} \Rightarrow \widetilde{T}_B(B) \equiv 0$ in $D_{11} \cup D_{12}$. If $\min(2a,1) > \epsilon \ge a$, then $n_1 = 0$ and $B \equiv 0$ in D_{11}. Suppose $n_1 \ge 1$ or, equivalently, $a > \epsilon$.

Claim: In D_{11}

1) $$\widetilde{T}_B^{\,n}(B)(x,y) \equiv 0 \text{ if } (y-a)^2 - (x-(a+\epsilon))^2 \le n\epsilon^2$$

for $0 \le n \le n_0+1$.

Proof of claim: If $n = 0$, $B(x,y) \equiv 0$ if $(x-y) \le \epsilon$ but $(y-a)^2-(x-(a+\epsilon))^2 = (x-y-\epsilon)(2a+\epsilon-x-y) \le 0$ iff $x-y \le \epsilon$, since $x+y < 2a+\epsilon$ in D_{11}. Suppose it is true for $0 \le n \le k \le n_0$

$$\widetilde{T}_B^{\,k}(B) \equiv 0 \text{ if } (y-a)^2 - (x-(a+\epsilon))^2 \le k\epsilon^2 \text{ in } D_{11}$$

2) $$\Gamma(\widetilde{T}_B^{\,k}(B)) \equiv 0 \text{ if } \max_{x \le \eta \le a} ((\eta-a)^2+(y-a)^2-(x-a)^2-(\eta-\epsilon-a)^2) \le k\epsilon^2$$

or $$\max_{x \le \eta \le a} ((y-a)^2-(x-a)^2+2\epsilon(\eta-a)) \le (k+1)\epsilon^2$$

or $$(y-a)^2-(x-a)^2 \le (k+1)\epsilon^2$$

3) $$\tilde{T}_B^{k+1}(B) = \int_y^x B(x,t)\Gamma^{(k)}(t,y)dt \equiv 0 \text{ if } (x-y) \le \epsilon$$

if $\epsilon < (x-y) \le \min(2a,1)$, then

4) $$\tilde{T}_B^{k+1}(B) = \int_y^{x-\epsilon} B(x,t)\Gamma^{(k)}(t,y)dt \equiv 0 \quad \text{if}$$

$$\max_{y\le t\le x-\epsilon} ((y-a)^2-(t-a)^2) \le \epsilon^2(k+1)$$

or $$(y-a)^2-(x-a-\epsilon)^2 \le \epsilon^2(k+1)$$ q.e.d.

$\therefore \quad \tilde{T}_B^{\,n}(B) \equiv 0$ in D_{11} if

5) $$\max_{0\le y\le a} (y-a)^2 \le n\epsilon^2 + \min_{0\le x\le a} (x-(a+\epsilon))^2$$

or $a^2 \le (n+1)\epsilon^2$ or $n \ge \dfrac{a^2}{\epsilon^2} - 1.$

$\therefore \quad \tilde{T}_B^{\,n}(B) \equiv 0$ in D_{11} if $n \ge n_1$.

Next consider $\tilde{T}_B^{\,n_1+j}(B)$ for $0 \le j \le n_2+ 1$.

6) Claim: $\tilde{T}_B^{\,n_1+j}(B) \equiv 0$ in D_{12} when $(x-y) \le (j+1)\epsilon$

Proof of claim: If $j = 0$, $\tilde{T}_B^{\,n_1+0}(B) \equiv 0$ if $(x-y) \le \epsilon$

Suppose it is true for $0 \le j = k < n_2+ 1$, then

7) $$\Gamma(\tilde{T}_B^{\,n_1+k}) \equiv 0 \text{ if } \max_{a\le\eta\le x}\left(\eta-a+\sqrt{(y-a)^2+(\eta-a)^2-(x-a)^2}\right) \le (k+1)\epsilon$$

or $(x-y) \le (k+1)\epsilon$

8) $$\tilde{T}_B^{\,n_1+k+1}(B) = \int_y^x B(x,t)\Gamma^{(n_1+k)}(t,y)dt$$

$$= \int_a^x B(x,t)\Gamma^{(n_1+k)}(t,y)dt$$

$$= 0 \quad \text{if } (x-a) \le \epsilon$$

if $x-a > \epsilon$, $\quad \tilde{T}_B^{n_1+k+1}(B) = \int_a^{x-\epsilon} B(x,t)\Gamma^{(n_1+k)}(t,y)dt = 0$

if $\quad \max\limits_{a \le t \le x-\epsilon} (t-y) \le (k+1)\epsilon$,

or if $\quad (x-y) \le (k+2)\epsilon$ q.e.d.

$\therefore \quad \tilde{T}_B^{n_1+m}(B) \equiv 0$ in $D_{11} \cup D_{12}$ if

9) $$\max_{(x,y)\varepsilon D_{12}} (x-y) \le (m+1)\epsilon .$$

Or if $$\min \frac{(2a,1)}{\epsilon} - 1 \le m$$

$\therefore \quad \tilde{T}_B^{n_0+1}(B) \equiv 0$ in $D_{11} \cup D_{12}$ q.e.d.

<u>III.7 Lemma</u> For any $j \ge 1$ and $\alpha \ge 2$, $\tilde{T}_B^{n_0+j}(B) \varepsilon \mu_\alpha$.

<u>Proof</u> By induction we shall show

1) $\tilde{T}_B^{n_0+j}(B) \equiv 0$ in $D_{11} \cup D_{12}$ and $\tilde{T}_B^{n_0+j}(B) \equiv 0$ if $(x-y) \le \epsilon$

2) $\tilde{T}_B^{n_0+j}(B)$ and $\frac{\partial}{\partial x} \tilde{T}_B^{n_0+j}(B)$ are absolutely continuous w.r. to x in $D_{22} \cup D_{21}$.

3) $\tilde{T}_B^{n_0+j}(B) \equiv 0$ in D_{21} if $x+y-2a \le \epsilon$

4) $\frac{\partial^2}{\partial x^2} \tilde{T}_B^{n_0+j}(B)$ is measurable on $D_{21} \cup D_{22}$ and

$$\operatorname*{ess\,sup}_{D_{21} \cup D_{22}} \left| \frac{\partial^2}{\partial x^2} \tilde{T}_B^{n_0+j}(B)(x,y) \right| < \infty$$

If we show the above, then $\tilde{T}_B^{n_0+j}(B) \varepsilon \mu_\alpha$ for any j and $\alpha > 2$.

For $j = 1$, we have shown $\tilde{T}_B^{n_0+1}(B) \equiv 0$ in $D_{11} \cup D_{12}$ and in D if $(x-y) \le \epsilon$. Expression 11 of Lemma III.4 gives us property 3). Absolute continuity, boundedness, and measurability follow in the now familiar fashion. We may go from j to j+1 in exactly the same manner.

III.8 Theorem $\exists$ an integer $n_3 \geq 0$ such that $\tilde{T}_B^{n_3+1}(B) \equiv 0$

Proof Recall, we have shown

$$\tilde{T}_B^{n_1}(B) \equiv 0 \quad \text{in } D_{11} \text{ and wherever } (x-y) \leq \epsilon ,$$

$$\Gamma(\tilde{T}_B^{n_1}(B)) \equiv 0 \quad \text{in } D_{12} \cup D_{21}$$

if
$$\max_{a \leq \eta \leq x} \left(\eta - a + \sqrt{(y-a)^2 + (\eta-a)^2 - (x-a)^2}\right) \leq \epsilon \quad \text{in } D_{12}$$

$$\max_{y \leq \xi \leq a} \left(a - \xi + \sqrt{(\xi-a)^2 + (x-a)^2 - (y-a)^2}\right) \leq \epsilon \quad \text{in } D_{21}$$

or if
$$(x-y) \leq \epsilon \quad \text{in } D_{12} \cup D_{21} .$$

By induction we may easily show that

$$\tilde{T}_B^{n_1+k}(B) \equiv 0 \quad \text{in } D_{12} \cup D_{21} \text{ whenever } (x-y) \leq (k+1)\epsilon$$

or
$$\tilde{T}_B^{n_1+k}(B) \equiv 0 \quad \text{in all of } D_{12} \cup D_{21} \text{ if } k \geq 1/\epsilon - 1$$

Let n_4 = smallest integer $\geq 1/\epsilon - 1$.

Consider $\tilde{T}_B^{n_1+n_4+j}(B)$ in D_{22} for $0 \leq j$.

If $1-a \leq \epsilon$, then $\tilde{T}_B^{n_1+n_4+j}(B) \equiv 0$ in D_{22}, and we are finished.

If not, Claim: if $j \geq 1$

1) $$\tilde{T}_B^{n_1+n_4+j}(B) \equiv 0 \quad \text{in } D_{22} \text{ if } (x-\epsilon-a)^2 - (y-a)^2 \leq j\epsilon^2$$

Proof of Claim: $\tilde{T}_B^{n_1+n_4}(B) \equiv 0$ if $(x-y) \leq \epsilon \Rightarrow \Gamma(\tilde{T}_B^{n_1+n_4}(B)) \equiv 0$ if

2) $$\max_{a \leq \xi \leq y} \left(a - \xi + \sqrt{(\xi-a)^2 + (x-a)^2 - (y-a)^2}\right) \leq \epsilon$$

or $(x-a)^2 - (y-a)^2 \leq \epsilon^2$

3) $$\tilde{T}_B^{n_1+n_4+1}(B) = \int_y^x B(x,t)\Gamma^{(n_1+n_4)}(t,y)\,dt \equiv 0$$

if $(x-y) \leq \epsilon$.

If $(x-y) > \epsilon$, then

4) $$\widetilde{T}_B^{\,n_1+n_4+1} = \int_y^{x-\epsilon} B(x,t)\Gamma^{(n_1+n_4)}(t,y)\,dt \equiv 0$$

if $$\max_{y\le t\le x-\epsilon} ((t-a)^2-(y-a)^2) \le \epsilon^2$$

or $$(x-a-\epsilon)^2 - (y-a)^2 \le \epsilon^2 .$$

Suppose it is true up to $j = k$ then

$$\Gamma(\widetilde{T}_B^{\,n_1+n_4+k}(B)) \equiv 0 \text{ in } D_{22} \quad \text{for}$$

5) $$\max_{a\le\xi\le y} \left[\left(\sqrt{(\xi-a)^2+(x-a)^2-(y-a)^2} - \epsilon\right)^2 - (\xi-a)^2\right] \le k\epsilon^2$$

or $$(x-a)^2-(y-a)^2- 2\epsilon\sqrt{(x-a)^2-(y-a)^2} - (k-1)\epsilon^2 \le 0$$

6) Let $$\sqrt{(x-a)^2-(y-a)^2} = z$$

7) Then we need $f(z) = z^2-2\epsilon z - (k-1)\,\epsilon^2 \le 0$

$$f(z) = 0 \quad \text{if } z = z_{\pm} = \epsilon(1\pm\sqrt{k})$$

8) $f(z)$ has a minimum at $z = \epsilon$

9) $\therefore$ $f(z) \le 0$ if $0 \le z \le \epsilon(1+\sqrt{k}\,)$

or if $(x-a)^2-(y-a)^2 \le (k+1)\epsilon^2 + 2\epsilon^2\sqrt{k}$.

10) or in D_{22} $\Gamma(\widetilde{T}_B^{\,n_1+n_4+k}(B)) \equiv 0$ if $(x-a)^2-(y-a)^2 \le (k+1)\epsilon^2$

then $$\int_y^{x-\epsilon} B(x,t)\Gamma^{(n_1+n_4+k)}(t,y)\,dt \equiv 0$$

if $$(x-a-\epsilon)^2-(y-a)^2 \le (k+1)\epsilon^2 .$$

Our claim is proven. It then follows that

11) $$\widetilde{T}_B^{\,n_1+n_4+j}(B) \equiv 0 \quad \text{in all of } D_{22} \quad \text{if}$$

$$\max_{a+\epsilon \le y+\epsilon < x \le 1} (x-a-\epsilon)^2 \le j\epsilon^2 + \min_{a\le y}(y-a)^2$$

or $$(1-(a+\epsilon))^2 \le j\epsilon^2$$

or $$j \ge \left(\frac{1-a}{\epsilon} - 1\right)^2$$

Let n_5 = smallest integer $\ge \left(\frac{1-a}{\epsilon} - 1\right)^2$ and $n_3 = n_1 + n_4 + n_5 - 1$.

We then have $$\tilde{T}_B^{\,n_1+n_4+n_5}(B) = \tilde{T}_B^{\,n_3+1}(B) \equiv 0 \text{ in } D$$ q.e.d.

III.9 Theorem For any complex number k and for any $B \in R_{a,\epsilon}$ it is true that

$$kQ_{a,\nu} + T_B \text{ is similar to } kQ_{a,\nu} \text{ on all } L_p[0,1] \quad 1 \le p \le \infty$$

Proof Let $$A_B = B - \frac{\tilde{T}_B(B)}{k} + \dots + \frac{(-1)^{n_3}}{k^{n_3}} \tilde{T}_B^{\,n_3}(B)$$

By Lemmas 5 and 7, $A_B \in \mu_{\alpha,a}$ for any $\alpha > 2$.

By Theorem 8, $1/k \;\; \tilde{T}_B(A_B) = -A_B + B$.

Thus $\Gamma(A_B)$ is a Volterra kernel on all $L_p[0,1]$ which satisfies

$$(kQ + T_B)(I - 1/k \; T_{\Gamma(A_B)}) = (I - 1/k \; T_{\Gamma(A_B)}) \, kQ$$

because $[Q, T_{\Gamma(A_B)}] = T_{A_B}$ by Theorem II.11.

III.10 Observation There is a certain asymmetry in our results. If a/ϵ is large, we need many more y derivatives in $D_{11} \cup D_{12}$ than the two necessary x derivatives in $D_{21} \cup D_{22}$. This observation leads us to the following.

III.11 Definition

$$\text{Let } v_1 = \text{smallest integer} \ge \max\left[\frac{(1-a)^2}{\epsilon^2} - 1,\ 0\right]$$

$$v_2 = \text{smallest integer} \ge \max\left[\min\left(2\,\frac{(1-a)}{\epsilon},\ 1/\epsilon\right) - 1, 0\right]$$

$$v_0 = \max[v_1 + v_2 - 1,\ 0]$$

for $0 < \epsilon < 1$ and $0 < a < 1$.

<u>III.12 Definition</u> Let $S_{a,\epsilon}$ consist of functions $B(x,y)$ defined on D such that

1) $B(x,y) \equiv 0$ if $(x-y) \le \epsilon$ a.e.

2) $\frac{\partial^j}{\partial y^j} B(x,y)$ is absolutely continuous w.r. to x a.e. in $D_{12} \cup D_{11}$ for $j = 0,1$.

3) $\frac{\partial^k}{\partial x^k} B(x,y)$ is absolutely continuous w.r. to x a.e. in $D_{21} \cup D_{22}$ for $k = 0,1,\ldots,2v_0+1$.

4) $B(x,y)$, $B_2(x,y)$, and $\frac{\partial^k}{\partial x^k} B(x,y)$ for $k = 1,2,\ldots,2v_0+1$ are all continuous functions of (x,y) in the closure of some neighborhood of the line $y = 2a-x$ contained in D. (If $a = 0$ or $a = 1$, this requirement is omitted).

5) $\frac{\partial^2}{\partial y^2} B(x,y)$ is measurable on $D_{11} \cup D_{12}$

$\frac{\partial^{2n_0+2}}{\partial x^{2n_0+2}} B(x,y)$ is measurable on $D_{21} \cup D_{22}$

$$\operatorname*{ess\ sup}_{D_{11} \cup D_{12}} \left| \frac{\partial^2}{\partial y^2} B(x,y) \right| < \infty$$

$$\operatorname*{ess\ sup}_{D_{21} \cup D_{22}} \left| \frac{\partial^{2n_0+2}}{\partial x^{2n_0+2}} B(x,y) \right| < \infty$$

$S_{a,\epsilon}$ is obviously a linear space.

III.13 Theorem For any non-zero complex number k and any $B \in S_{a,\epsilon}$, $kQ_{a,\nu} + T_B$ is similar to $kQ_{a,\nu}$ on all the Lebesgue spaces $L_p[0,1]$.

Proof $B(x,y) \in S_{a,\epsilon}$ iff $B(1-y, 1-x) \in R_{1-a,\epsilon}$. Hence, by Theorem 9

$$-kQ_{1-a,1-\nu} + T_{B(1-y,1-x)} \text{ is similar to } -kQ_{1-a,1-\nu}$$

on all $L_p[0,1]$. Then the result follows by Lemma I.6 of the preceding paper.

III.14 Observation If $a = 0$, then $n_0 = 0$ for any ϵ. In contrast, if $a > 0$, then $\lim_{\epsilon \to 0} n_0(a,\epsilon) = +\infty$. This observation leads us to expect that we may perhaps obtain $kQ_{0,\nu} + T_B$ similar to $kQ_{0,\nu}$ for some functions, $B(x,y)$, for which $B(x,y)$ need not vanish in a neighborhood of the diagonal.

An identical argument follows for $a = 1$ and $v_0(a,\epsilon)$.

III.15 Lemma $L_{\alpha,0}$ is a Banach space if we let

$$\|B\| = |B|^{\alpha,0} = \operatorname*{ess\,sup}_D |(x-y)^{1-\alpha} B_{11}(x,y)|$$

and identify kernels which differ on a set of measure zero.

Proof Trivial consequence of

$$B(x,y) = \int_y^x (x-t)B_{11}(t,y)dt \quad \text{a.e.}$$

III.16 Lemma If A and $B \in L_{\alpha,0}$ for $\alpha > \frac{1}{2}$ then $\tilde{T}_B(A) \in L_{\alpha,0}$ and

$$\|\tilde{T}_B(A)\| \le C(\alpha,\nu) \|B\| \|A\|$$

for $C(\alpha,\nu) = \beta(\alpha, \min(\alpha/2, \alpha-\frac{1}{2}))d_{11}(\alpha,1,1-\nu)$.

Proof Trivial consequence of Lemmas I.6 and II.9. The following two Theorems are valid on all $L_p[0,1]$ $1 \le p \le \infty$.

III.17 Theorem $kQ_{0,\nu}$ is similar to $kQ_{0,\nu} + T_B$ for any non-zero complex number k and any $\alpha > \frac{1}{2}$ such that

$$|B|^{\alpha,0} < \frac{|k|}{C(\alpha,\nu)}$$

Proof By Lemma 16

$$\frac{|\tilde{T}_B(A)|^{\alpha,0}}{|k|} \leq \frac{|B|^{\alpha,0}C(\alpha,\nu)|A|^{\alpha,0}}{|k|} \leq \theta|A|^{\alpha,0}$$

where $\theta < 1$.

Hence $(I + 1/k\ \tilde{T}_B)A = B$ possesses a unique solution A_B in $\mu_{\alpha,0}$,

$$A_B = (I + 1/k\ \tilde{T}_B)^{-1} B$$

and

$$(kQ_{0,\nu} + T_B)(I - 1/k\ T_{\Gamma(A_B)}) = (I - 1/k\ T_{\Gamma(A_B)}) \cdot kQ_{0,\nu} .$$

where $T_{\Gamma(A_B)}$ is a Volterra operator on all $L_p[0,1]$.

III.18 Lemma $kQ_{1,\nu}$ is similar to $kQ_{1,\nu} + T_B$ for any complex number k and for any B in $L_{\alpha,1}$ with $\alpha > \frac{1}{2}$ such that

$$|B|^{\alpha,1} < \frac{|k|}{C(\alpha,1-\nu)}$$

Proof Direct consequence of Theorem 17 and Lemma I.6 of the preceding paper.

III.19 Definition If $\alpha > 2$, $0 < a < 1$, let $N_{\alpha,a}$ consist of all B in $\mu_{\alpha,a}$ such that $B(x,y) \equiv 0$ a.e. in $D_{11}^{(a)} \cup D_{22}^{(a)}$

III.20 Theorem On all $L_p[0,1]$, $1 \leq p \leq \infty$, $kQ_{a,\nu} + T_B$ is similar to $kQ_{a,\nu}$ for any complex non-zero k and any B in $N_{\alpha,a}$.

Proof Let $A_B = B$. By Lemma II.3, $\Gamma(A_B) \equiv 0$ a.e. in $D_{11} \cup D_{22}$.

It is easily seen that

$$\tilde{T}_B(A_B) = B*\Gamma(B) \equiv 0 \quad \text{a.e.}$$

Thus, by Theorems II.12 and II.8

$$(kQ_{a,\nu} + T_B)(I - 1/k\, T_{\Gamma(B)}) = (I - 1/k\, T_{\Gamma(B)})\, kQ_{a,\nu}$$

and $T_{\Gamma(B)}$ is a Volterra integral operator on all $L_p[0,1]$.

III.21 Lemma Let α and a be real numbers such that $\alpha > \frac{1}{2}$ and $0 < a < 1$. Then $L_{\alpha,a}$ is a Banach space, if we define

$$\|A\| = |A|^{\alpha,a} = \max[\operatorname*{ess\,sup}_{D_{11}} |(x-y)^{1-\alpha} A_{22}(x,y)|, \operatorname*{ess\,sup}_{D_{22}} |(x-y)^{1-\alpha} A_{11}(x,y)|]$$

and identify kernels, A, which differ on a set of measure zero.

Proof $\quad A(x,y) = \int_y^x (x-t) A_{11}(t,y)\,dt$ in D_{22} a.e.

$$A(x,y) = \int_y^x A_{22}(x,t)(t-y)\,dt \quad \text{in } D_{11} \quad \text{a.e.}$$

$$A(x,y) \equiv 0 \quad \text{in } D_{12} \cup D_{21} \quad \text{a.e.}$$

The result follows trivially.

III.22 Lemma If $\alpha > 2-\nu$ and A and B are in $L_{\alpha,a}$, then $\tilde{T}_B(A) \in L_{\alpha,a}$ and

(1) $$\left\| \frac{\partial^2}{\partial x^2} \tilde{T}_B(A) \right\|_{\alpha,a,D_{22}} \le \beta(\alpha,\alpha/2)\, d_{11}(\alpha,1-a,1-\nu) \|A_{11}\|_{\alpha,a,D_{22}} \|B_{11}\|_{\alpha,a,D_{22}}$$

(2) $$\left\| \frac{\partial^2}{\partial y^2} \tilde{T}_B(A) \right\|_{\alpha,a,D_{11}} \le b(\alpha,a,\nu) \|A_{22}\|_{\alpha,a,D_{11}} \|B_{22}\|_{\alpha,a,D_{11}}$$

$$b(\alpha,a,\nu) = \beta(\alpha,\alpha)\left[\frac{1}{\alpha} + \frac{(1-\nu)}{\alpha(\alpha+1)}\right] + \beta(\alpha,\alpha/2 + \nu/2 - 1)\, \frac{(2a)^{\alpha/2 + 1 - \nu/2}}{2-\nu}$$

$$\left[1 + \frac{3-2\nu}{\alpha} + \frac{(1-\nu)(3-\nu)}{\alpha(\alpha+1)}\right]\left[1 + \frac{(3-2\nu)2}{\alpha} + \frac{2(1-\nu)(3-\nu)}{\alpha(\alpha+1)}\right]$$

Proof We first show $\tilde{T}_B(A) \equiv 0$ in $D_{12} \cup D_{21}$ a.e.

In D_{22}, $$\frac{\partial^2}{\partial x^2} \tilde{T}_B(A) = \int_y^x B_{11}(x,t)\Gamma_{(22)}(A)(t,y)\,dt \quad \text{a.e.}$$

Part (1) then follows simply.

In D_{11} we may integrate by parts and obtain

$$\tilde{T}_B(A) = -\int_y^x (y-a)^{1-\nu} \left[\frac{B_{22}(x,t)}{(t-a)^{2-\nu}} - (3-2\nu) \frac{B_2(x,t)}{(t-a)^{3-\nu}} + (1-\nu)(3-\nu) \frac{B(x,t)}{(t-a)^{4-\nu}} \right] dt$$

$$\int_a^t (\eta-a)^{1-\nu} d\eta \frac{A\left(\eta, a-\sqrt{(y-a)^2+(\eta-a)^2-(t-a)^2}\right)}{\left(-\sqrt{(y-a)^2+(\eta-a)^2-(t-a)^2}\right)^{1-\nu}}$$

$$-\int_y^x \left[\frac{B_2(x,t)}{(t-a)} - (1-\nu) \frac{B(x,t)}{(t-a)^2} \right] A(t,y)\,dt \qquad \text{a.e.}$$

Necessary properties of the y derivatives of $\tilde{T}_B(A)$ in D_{11} then follow simply.

The estimate of $\frac{\partial^2}{\partial y^2} \tilde{T}_B(A)$ in D_{11} is also obtained in a straightforward manner from the above equality.

III.23 Theorem Let

(1) k be any non-zero complex number

(2) $0 < a < 1$,

(3) $\alpha > \min(2-\nu, 1+\nu)$

(4) $B \in L_{\alpha,a}$

Then $kQ_{a,\nu} + T_B$ is similar to $kQ_{a,\nu}$ on all Lebesgue spaces $L_p[0,1]$, $1 \le p \le \infty$, if

(a) $\alpha \geq 2-\nu$ and $\|B_{11}\|_{\alpha,a,D_{22}} < \frac{|k|}{\beta(\alpha,\alpha/2)\, d_{11}(\alpha,1-a,1-\nu)}$

$\|B_{22}\|_{\alpha,a,D_{11}} < \frac{|k|}{b(\alpha,a,\nu)}$

or

(b) $\alpha > 1+\nu$ and $\|B_{11}\|_{\alpha,a,D_{22}} < \frac{|k|}{b(\alpha,1-a,1-\nu)}$

$$\|B_{22}\|_{\alpha,a,D_{11}} < \frac{|k|}{\beta(\alpha,\alpha/2)d_{11}(\alpha,a,\nu)}$$

<u>Proof</u> Part (a) follows immediately from Lemma 22.

Thus $kQ_{1-a,1-\nu} - T_{B(1-y,1-x)}$ is similar to $kQ_{1-a,1-\nu}$ on all $L_p[0,1]$ if $\widetilde{B}(x,y) = B(1-y,1-x)$ satisfies

$$\|\widetilde{B}_{11}\|_{\alpha,1-a,D_{22}} < \frac{|k|}{\beta(\alpha,\alpha/2)d_{11}(\alpha,a,\nu)}$$

$$\|\widetilde{B}_{22}\|_{\alpha,1-a,D_{11}} < \frac{|k|}{b(\alpha,1-a,1-\nu)}$$

and $\alpha > 1+\nu$ and $\widetilde{B} \in L_{\alpha,1-a}$ Part (b) then is a consequence of Lemma I.6 of the preceding paper.

<u>III.24 Summary</u> The following statements hold on all Lebesgue spaces $L_p[0,1]$, $(1 \leq p \leq \infty)$ and for all non-zero complex numbers k.

1) $kQ_{a,\nu} + T_B$ is similar to $kQ_{a,\nu}$ for any B in $R_{a,\epsilon}$ (Theorem III.9). $R_{a,\epsilon}$ is defined in III.4.

2) $kQ_{a,\nu} + T_B$ is similar to $kQ_{a,\nu}$ for any B in $S_{a,\epsilon}$ (Theorem III.13). $S_{a,\epsilon}$ is defined in III.12.

3) $kQ_{0,\nu} + T_B$ is similar to $kQ_{0,\nu}$ for any B in $L_{\alpha,0}$ $(\alpha > \frac{1}{2})$ such that $|B|^{\alpha,0} < |k|/C(\alpha,\nu)$ (Theorem III.17). $L_{\alpha,0}$ is defined in II.10.

4) $kQ_{1,\nu} + T_B$ is similar to $kQ_{1,\nu}$ for any B in $L_{\alpha,1}$ $(\alpha > \frac{1}{2})$, such that $|B|^{\alpha,1} < \dfrac{|k|}{C(\alpha,1-\nu)}$ (Theorem III.18). $L_{\alpha,1}$ is defined in II.10.

5) $kQ_{a,\nu} + T_B$ is similar to $kQ_{a,\nu}$ for any B in $N_{\alpha,a}$ for $\alpha > 2$. (Theorem III.20). $N_{\alpha,a}$ is defined in III.19.

6) $kQ_{a,\nu} + T_B$ is similar to $kQ_{a,\nu}$ for B in $L_{\alpha,a}$ $0 < a < 1$, if either

a) $\alpha > 2-\nu$ and $\|B_{11}\|_{\alpha,a,D_{22}} < \dfrac{|k|}{\not{b}(\alpha,\alpha/2)\,d_{11}(\alpha,1-a,1-\nu)}$

and $\|B_{22}\|_{\alpha,a,D_{11}} < \dfrac{|k|}{b(\alpha,a,\nu)}$

or

b) $\alpha > 1+\nu$ and $\|B_{11}\|_{\alpha,a,D_{22}} < \dfrac{|k|}{b(\alpha,1-a,1-\nu)}$

and $\|B_{22}\|_{\alpha,a,D_{11}} < \dfrac{|k|}{\not{b}(\alpha,\alpha/2)\,d_{11}(\alpha,a,\nu)}$

(Theorem III.23). $L_{\alpha,a}$ is defined in II.10.

IV. EXTENSION OF THE RESULTS

When we used Friedrichs' method to prove that kQ is similar to a Volterra operator T_G, we examined functions G(x,y) which were a perturbation of $k(x-a)^{\nu}(y-a)^{1-\nu}$. In this section we shall consider Volterra operators T_G such that T_G may be reduced by means of elementary similarity transformations to perturbations of kQ

$$V^{-1}\,T_G V = kQ + T_B$$

<u>IV.1 Definition</u> Let S(x) map $[0,1] \leftrightarrow [0,1]$ one to one onto with inverse map $S^{-1}(x) = R(x)$ and let $S'(x) > 0$ and S(x) be absolutely continuous. Then $S_S f = f(S(x))$ is a bounded linear map of $L_p[0,1] \to L_p[0,1]$ with inverse $S_S^{-1} f = S_R f = f(R(x))$.

<u>IV.2 Definition</u> Let $\ell(x)$ be measurable on $[0,1]$ with the property that

$$\operatorname*{ess\,sup}_{0\le x\le 1}\left(|\ell(x)| + \frac{1}{|\ell(x)|}\right) < \infty\ .$$

Then $M_\ell f = \ell(x)\cdot f(x)$ is a bounded linear map of $L_p[0,1] \to L_p[0,1]$ with inverse $M_{\frac{1}{\ell}} = M_\ell^{-1}$. $M_{\frac{1}{\ell}} f = \dfrac{f(x)}{\ell(x)}$.

IV.3 Lemma If T_A and T_B are Volterra integral operators on $L_p[0,1]$ and $T_A = M_{\frac{1}{\ell}} S_R T_B S_S M_\ell$, then

$$A(x,y) = B(R(x),R(y))R'(y) \frac{\ell(y)}{\ell(x)} \quad \text{a.e.}$$

Proof Trivial.

IV.4 Lemma If B is of rank 1 with diagonal zero at b and B is continuous w.r. to x in $D_{22}^{(b)}$, continuous w.r. to y in $D_{11}^{(b)}$, measurable and essentially bounded on D, then ∃ a unique pair of constants, k and a, satisfying

1) $k > 0,\ 0 \le a \le 1$

2) $\int_0^b B(t,t)dt = k\int_0^a (t-a)dt = -k/2\ a^2 = C_1$

3) $\int_0^1 B(t,t)dt = k\int_0^1 (t-a)dt = k/2\ (1-2a) = C_2$

Proof Straightforward calculation yields

$$\left.\begin{array}{ll} k = 2C_2 & \text{if } C_1 = 0 \\ k = -2C_1/a^2 & \text{if } C_1 < 0 \end{array}\right\} \quad \left\{\begin{array}{ll} a = \frac{C_1}{C_2} + \sqrt{C_1^2/C_2^2 - C_1/C_2} & \text{if } C_2 > 0 \\ a = \frac{1}{2} & \text{if } C_2 = 0 \\ a = C_1/C_2 - \sqrt{C_1^2/C_2^2 - C_1/C_2} & \text{if } C_2 < 0 \end{array}\right.$$

IV.5 Lemma If

(1) B(x,y) is of rank 1 and has certain herein unspecified continuity conditions on D.

(2) k and a are chosen as in IV.4

(3) $$\frac{[B_1(x,x)]_{x=b}}{\lim_{x\to b}[d/dx\ B(x,x)]} = \nu \text{ and/or } \frac{[B_2(x,x)]_{x=b}}{\lim_{x\to b}[d/dx\ B(x,x)]} = 1-\nu$$

where $0 \le \nu \le 1$. Then $M_{\frac{1}{\ell}} S_R T_B S_S M_\ell = kQ_{a,\nu} + T_G$, where G has the properties

(4) if $x \le a$, $\lim_{y \to x} G(x,y) = \lim_{y \to x} G_2(x,y) = 0$ a.e.

(5) if $y \ge a$, $\lim_{x \to y} G(x,y) = \lim_{x \to y} G_1(x,y) = 0$ a.e.

and

(6) $S(x) = a - \sqrt{a^2 + 2/k \int_0^x B(t,t)dt}$ if $x < b$

$S(x) = a + \sqrt{a^2 + 2/k \int_0^x B(t,t)dt}$ if $x \ge b$

(7) $R(x) = S^{-1}(x)$ satisfies

$$\int_0^{R(x)} B(t,t)dt = \int_0^x k(t-a)dt \quad \text{such that}$$

if $x < a$, $R(x) < b$

if $x = a$, $R(a) = b$

if $x > a$, $R(x) > b$

(8) $$\ell(x) = \exp \int_a^x \frac{B_1(R(t),R(t))[R'(t)]^2 - k\nu}{k(t-a)} dt \quad \text{if } x \ge a$$

$$\ell(x) = \exp \int_a^x \frac{[\left((d/dy\ B(y,y)\right)_{y=R(t)} - B_2(R(t),R(t))][R'(t)]^2 - k\nu}{k(t-a)} dt$$

if $x < a$.

Proof Straightforward calculation if we assume sufficient continuity criteria for $B(x,y)$.

IV.6 Remark We shall use the transformations of IV.5 to obtain six results which correspond to the six categories of Summary III.24. We shall use the method outlined at the beginning of this section. Each time we use the expression $(f)^{\mu}$ in what follows, either (a) $0 \le \arg f < 2\pi$ or (b) $-2\pi < \arg f \le 0$. We shall use (a) when we refer to a result concerning $Q^+_{a,\nu}$, (b) in a result involving $Q^-_{a,\nu}$. (See preceding paper I.1, I.2 and I.3). In the following six theorems, a, k, ν, b, $S(x)$, $\ell(x)$, will be defined as in IV.5 and IV.4. The quantities n_0 and v_0 will be defined as in III.3 and III.11 respectively. Similarity will be valid on all $L_p[0,1]$, $1 \le p \le \infty$, and all $B(x,y)$ will be of rank one.

<u>IV.7 Theorem</u> Let $B(x,y)$ be defined on D such that is satisfies (1) to (5). Then T_B is similar to $kQ_{a,\nu}$

(1) $B(x,y) = (x-b)^{\nu}(y-b)^{1-\nu}f(x)g(y) + H(x,y)$ such that $H(x,y) \equiv 0$ if $S(x)-S(y) \le \epsilon$ a.e. and $f(x) > 0$ and $g(x) > 0$ on the interval $[0,1]$.

(2) $f(x)g(x) \in C^{2n_0+1}[0,1]$

$\left(\frac{\partial}{\partial x}\right)^{2n_0+2}(f(x)g(x))$ is continuous in a neighborhood of $x = b$.

$f(x)g'(x) \in C^{2n_0+1}[0,1]$

$\left(\frac{\partial}{\partial x}\right)^{2n_0+2}(f(x)g'(x))$ is continuous in a neighborhood of $x = b$

$f(x)$ and $g(x)$ are continuous on the interval $[0,1]$.

(3) Let $D^{(a)}_{ij,S} = \{(x,y) \in D \,|\, (S(x),S(y)) \in D^{(a)}_{ij}\}$

$\frac{\partial^k}{\partial x^k} H(x,y)$ is absolutely continuous w.r. to x in $D_{21,S} \cup D_{22,S}$ for $k = 0,1$ a.e.

$\frac{\partial^j}{\partial y^j} H(x,y)$ is absolutely continuous w.r. to y in $D_{11,S} \cup D_{12,S}$ for $j = 0,1,\ldots,2n_0+1$ a.e.

(4) $\frac{\partial^k}{\partial y^k} H(x,y)$ for $k = 0,1,2,\ldots,2n_0+1$ and $H_1(x,y)$ are all continuous functions of (x,y) in the closure of some neighborhood of the curve $C = \{(x,y) \,|\, S(x)+S(y)=2a\} \cap D$.

(5) $\frac{\partial^{2n_0+2}}{\partial y^{2n_0+2}} H(x,y)$ is measurable on $D_{11,S} \cup D_{12,S}$

$\frac{\partial^2}{\partial x^2} H(x,y)$ is measurable on $D_{21,S} \cup D_{22,S}$

$$\operatorname*{ess\ sup}_{D_{21,S} \cup D_{22,S}} \left| \frac{\partial^2}{\partial x^2} H(x,y) \right| < \infty$$

$$\operatorname*{ess\ sup}_{D_{11,S} \cup D_{12,S}} \left| \frac{\partial^{2n_0+2}}{\partial y^{2n_0+2}} H(x,y) \right| < \infty$$

<u>IV.8 Theorem</u> Let $B(x,y)$ be defined on D such that it satisfies (1) to (5). Then T_B is similar to $kQ_{a,\nu}$

(1) $B(x,y) = (x-b)^{\nu}(y-b)^{1-\nu}f(x)g(y) + H(x,y)$ such that $H(x,y) \equiv 0$ if $S(x)-S(y) \le \epsilon$ a.e. and $f(x) > 0$ and $g(x) > 0$ on the interval $[0,1]$.

(2) $f(x)g(x) \in C^{\max(2v_0,1)}[0,1]$

$\left(\frac{\partial}{\partial x}\right)^{\max(2v_0+1,2)} (f(x)\cdot g(x))$ is continuous in a neighborhood of $x = b$

$f(x)g'(x) \in C^{2v_0+1}[0,1]$

$\left(\frac{\partial}{\partial x}\right)^{2v_0+2} (f(x)g'(x)$ is continuous in a neighborhood of $x = b$

$f(x)$ and $g(x)$ are continuous on the interval $[0,1]$.

(3) $\frac{\partial^k}{\partial x^k} H(x,y)$ is absolutely continuous w.r. to x in $D_{21,S} \cup D_{22,S}$ for $k = 0,1,\dots,2v_0+1$ a.e.

$\frac{\partial^j}{\partial y^j} H(x,y)$ is absolutely continuous w.r. to y in $D_{11,S} \cup D_{12,s}$ for $j = 0,1,$ a.e.

(4) $\frac{\partial^k}{\partial x^k} H(x,y)$ for $k = 0,1,2,\ldots,2v_0+1$ and $H_2(x,y)$ are all continuous functions of (x,y) in the closure of some neighborhood of the curve $C = \{(x,y) \mid S(x)+S(y) = 2a\} \cap D$.

(5) $\frac{\partial^{2v_0+2}}{\partial x^{2v_0+2}} H(x,y)$ is measurable on $D_{21,S} \cup D_{22,S}$

$\frac{\partial^2}{\partial y^2} H(x,y)$ is measurable on $D_{11,S} \cup D_{12,S}$

$$\operatorname*{ess\,sup}_{D_{21,S} \cup D_{22,S}} \left| \frac{\partial^{2v_0+2}}{\partial x^{2v_0+2}} H(x,y) \right| < \infty$$

$$\operatorname*{ess\,sup}_{D_{11,S} \cup D_{12,S}} \left| \frac{\partial^2}{\partial y^2} H(x,y) \right| < \infty$$

IV.9 Theorem Let $b = 0$ and let $B(x,y)$ satisfy (1) to (4). Then T_B is similar to $kQ_{0,\nu}$.

(1) $B(x,y) = x^{\nu} y^{1-\nu} f(x)g(y) + H(x,y)$ with $f(x) > 0$ and $g(x) > 0$ on the interval $[0,1]$ and

$$\lim_{x \to y} H(x,y) = \lim_{x \to y} H_1(x,y) = 0 \quad \text{a.e.}$$

(2) $f(x)g(x) \in C^1[0,1]$

$\frac{\partial^2}{\partial x^2}(f(x)g(x))$ is continuous in a neighborhood of $x = 0$

$f(x)g'(x) \in C^1[0,1]$

$\frac{\partial^2}{\partial x^2}(f(x)g'(x))$ is continuous in a neighborhood of $x = 0$

$f(x)$ and $g(x)$ are continuous on the interval $[0,1]$.

(3) $\frac{\partial^k}{\partial x^k} H(x,y)$ is absolutely continuous w.r. to x a.e. for $k = 0,1$.

(4) $\frac{\partial^2}{\partial x^2} H(x,y)$ is measurable and $\exists\ \alpha > \frac{1}{2}$ such that

$$\operatorname*{ess\,sup}_{D} \left| (x-y)^{1-\alpha} \frac{\partial^2}{\partial x^2} \left[H(R(x),R(y))R'(x) \frac{\ell(y)}{\ell(x)} \right] \right| < \frac{|k|}{C(\alpha,\nu)}$$

<u>IV.10 Theorem</u> Let b = 1 and let B(x,y) satisfy (1) to (4). Then T_B is similar to $kQ_{1,\nu}$.

(1) $B(x,y) = (x-1)^{\nu}(y-1)^{1-\nu}f(x)g(y) + H(x,y)$ with $f(x) > 0$ and $g(x) > 0$ on the interval $[0,1]$ and

$$\lim_{y\to x} H(x,y) = \lim_{y\to x} H_2(x,y) = 0 \quad \text{a.e.}$$

(2) $f(x)g(x) \in C^1[0,1]$

$\frac{\partial^2}{\partial x^2}(f(x)g(x))$ is continuous in a neighborhood of x = 1.

$f(x)g'(x) \in C^1[0,1]$

$\frac{\partial^2}{\partial x^2}(f(x)g'(x))$ is continuous in a neighborhood of x = 1.

(3) $\frac{\partial^k}{\partial y^k} H(x,y)$ is absolutely continuous w.r. to y a.e.

for k = 0,1.

(4) $\frac{\partial^2}{\partial y^2} H(x,y)$ is measurable and $\exists\ \alpha > \frac{1}{2}$ such that

$$\operatorname*{ess\,sup}_{D} \left| (x-y)^{1-\alpha} \frac{\partial^2}{\partial y^2} H(R(x),R(y))R'(x) \frac{\ell(y)}{\ell(x)} \right| < \frac{|k|}{C(\alpha,1-\nu)}$$

<u>IV.11 Theorem</u> Let B(x,y) be defined on D and satisfy (1) to (4). Then $kQ_{a,\nu} + T_B$ is similar to $kQ_{a,\nu}$

(1) $B(x,y) = f(x)g(y)(x-b)^{\nu}(y-b)^{1-\nu} + H(x,y)$ such that $H \equiv 0$ in $D_{11,S} \cup D_{22,S}$ a.e.

(2) $f(x)g(x) \in C^1[0,1]$

$\frac{\partial^2}{\partial x^2}(f(x)g(x))$ is continuous in a neighborhood of $x = b$

$f(x)g'(x) \in C^1[0,1]$

$\frac{\partial^2}{\partial x^2}(f(x)g'(x))$ is continuous in a neighborhood of $x = b$.

(3) $\frac{\partial^k}{\partial x^k} H(x,y)$ is absolutely continuous w.r. to x in $D_{21,S}$ for $k = 0,1$ a.e.

$\frac{\partial^j}{\partial y^j} H(x,y)$ is absolutely continuous w.r. to y in $D_{12,S}$ for $j = 0,1$ a.e.

(4) $H(x,y)$, $H_1(x,y)$ and $H_2(x,y)$ are all continuous functions of (x,y) in the closure of some neighborhood of the curve

$$C = \{(x,y) \mid S(x) + S(y) = 2a\} \cap D$$

(5) $\frac{\partial^2}{\partial x^2} H(x,y)$ is measurable on $D_{21,S}$

$\frac{\partial^2}{\partial y^2} H(x,y)$ is measurable on $D_{12,S}$

$\exists\ \alpha > 2$ such that $|\hat{H}|^{\alpha,a} < \infty$ where $\hat{H} = H(R(x),R(y))R'(y)\,\frac{\ell(y)}{\ell(x)}$

<u>IV.12 Theorem</u> Let $B(x,y)$ be defined on D and satisfy (1) to (4). Then $kQ_{a,\nu} + T_B$ is similar to $kQ_{a,\nu}$.

(1) $B(x,y) = f(x)g(y)(x-b)^{\nu}(y-b)^{1-\nu} + H(x,y)$ such that $H \equiv 0$ in $D_{21,S} \cup D_{12,S}$ a.e. and

(a) if $x < b$, then $\lim_{y \to x} H(x,y) = \lim_{y \to x} H_2(x,y) = 0$ a.e.

(b) if $y > b$, then $\lim_{x \to y} H(x,y) = \lim_{x \to y} H_1(x,y) = 0$ a.e.

(2) $f(x)g(x) \in C^1[0,1]$

$\frac{\partial^2}{\partial x^2}(f(x)g(x))$ is continuous in a neighborhood of $x = b$.

$f(x)g'(x) \in C^1[0,1]$

$\frac{\partial^2}{\partial x^2}(f(x)g'(x))$ is continuous in a neighborhood of $x = b$.

(3) $\frac{\partial^k}{\partial x^k} H(x,y)$ is absolutely continuous w.r. to x in $D_{22,S}$ for $k = 0,1$ a.e.

$\frac{\partial^j}{\partial y^j} H(x,y)$ is absolutely continuous w.r. to y in $D_{11,S}$ for $j = 0,1$ a.e.

(4) $\frac{\partial^2}{\partial x^2} H(x,y)$ is measurable in $D_{22,S}$

$\frac{\partial^2}{\partial y^2} H(x,y)$ is measurable in $D_{11,S}$

$\exists\ \alpha > \min(1+\nu, 2-\nu)$ such that either

(a) $\alpha > 2-\nu$ and $\|\hat{H}_{11}\|_{\alpha,a,D_{22}} < \frac{|k|}{B(\alpha,\alpha/2)\,d_{11}(\alpha,1-a,1-\nu)}$

and $\|\hat{H}_{22}\|_{\alpha,a,D_{11}} < \frac{|k|}{b(\alpha,a,\nu)}$

or (b) $\alpha > 1+\nu$ and $\|\hat{H}_{11}\|_{\alpha,a,D_{22}} < \frac{|k|}{b(\alpha,1-a,1-\nu)}$

and $\|\hat{H}_{22}\|_{\alpha,a,D_{11}} < \frac{|k|}{B(\alpha,\alpha/2)\,d_{11}(\alpha,a,\nu)}$

V. REMARKS AND SUMMARY

Our results may be briefly summarized as follows:

If T_A and T_B are of rank one with diagonal zeros at a and b respectively, then T_A is similar to T_B on each $L_p[0,1]$ if

(1) $$\int_0^1 A(t,t)dt = \int_0^1 B(t,t)dt$$

(2) $$\int_0^a A(t,t)dt = \int_0^b B(t,t)dt$$

(3) $$0 \leq \lim_{y\to a} \frac{A_1(y,y)}{[d/dy\ A(y,y)]} = \lim_{y\to b} \frac{B_1(y,y)}{[d/dy\ B(y,y)]} \leq 1$$

assuming A and B both have certain smoothness properties.

An approximate converse to these results is obtained from the preceding paper.

The corresponding similarity invariant for operators of order α was

(4) $$\int_0^1 [A(t,t)]^{1/\alpha}\, dt$$

The results appear to be generalizable to Volterra operators having numerous zeros of various orders on the diagonal.

The main difficulty in generalizing the results comes in solving the commutator equation and indeed in choosing the proper canonical form so that this may be done.

In a future paper the author will present a general method for obtaining the canonical form and solving the corresponding commutator equation.

ACKNOWLEDGMENT

The author wishes to thank Professor J. T. Schwartz for his guidance in this work, which is based upon a doctoral dissertation at New York University.

BIBLIOGRAPHY

1. Dunford, N. A Survey of the Theory of Spectral Operators. A. M. S. Bull. Vol. 64, (1958), pp. 217-274.

2. Dunford, N. and J. T. Schwartz. Linear Operators, Part I: General Theory. Interscience, N. Y. (1958).

3. Dupras, A. Doctoral Thesis, New York University (1965).

4. Freeman, J. M. Volterra Operators Similar to J. $J : f \to \int_0^x f(t)dt$. Trans. Amer. Math. Soc. (4) Vol. 116 (1965), pp. 181-192.

5. Friedrichs, K. O. Perturbations of Continuous Spectra. Comm. Pure Appl. Math., Vol. 1 (1948), pp. 361-406.

6. Kalisch, G. K. On Similarity, Reducing Manifolds, and Unitary Equivalence of Certain Volterra Operators. Ann. of Math. (2), Vol. 66 (1957), pp. 481-494.

7. Kalisch, G. K. On Isometric Equivalence of Certain Volterra Operators. Proc. A. M. S., Vol. 12 (1961), pp. 93-98.

8. Kalisch, G. K. On Similarity Invariants of Certain Volterra Operators on L_p. Pacific J. Math., Vol. 11, (1961), pp. 247-252.

9. Schwartz, J. T. Some Non-Self Adjoint Operators. Comm. Pure Appl. Math., Vol. 8 (1960), pp. 609-639.

10. Volterra, V. Theory of Functionals and of Integral and Integro-Differential Equations, Dover, New York (1959).

11. Volterra, V. and J. Pérès. Lecons sur la Composition et les Fonctions Permutables. Gauthiers-Villars, Paris (1925).